A

TRUE DESCRIPTION

OF THE

LAKE SUPERIOR COUNTRY;

ITS RIVERS, COASTS, BAYS, HARBOURS, ISLANDS, AND COMMERCE.

WITH

BAYFIELD'S CHART;

[Showing the Boundary Line as Established by Joint Commission.]

ALSO A MINUTE ACCOUNT OF THE

COPPER MINES

AND

WORKING COMPANIES.

ACCOMPANIED BY

A MAP OF THE MINERAL REGIONS;

SHOWING, BY THEIR NO. AND PLACE, ALL THE DIFFERENT LOCATIONS:

AND CONTAINING

A CONCISE MODE OF ASSAYING, TREATING, SMELTING, AND REFINING COPPER ORES.

BY

JOHN R. ST. JOHN.

NEW YORK:

WILLIAM H. GRAHAM, TRIBUNE BUILDINGS.

1846.

ADVERTISEMENT.

To arrive at a reasonable accuracy in describing a new and partially explored country, the explorations of which have been by very many different individuals, a mode of giving the description must be adopted corresponding to that by which a conflict between two armies is described. No one person having been able to see all the different points to be described, the statements of various persons must be taken, in order to a knowledge of the whole.

In preparing this work, therefore, I have had recourse to every available source of information, scrutinizing and comparing the different statements. And to the end of my principal object, that of presenting, in a condensed form, the disseminated facts which great national interests require the public to be in possession of, I make no apology, but would give full credit for the copious drafts and condensations from the Reports of Officers of State and general governments, as well as the personal communications of many intelligent gentlemen, for whose contributions I am under many obligations.

Facts and information which I have been enabled to obtain only by perseverance, toil, and personal observation, are joined therewith, together with deductions sifted from concurring accounts by old *voyageurs*, half-breeds and Indians, delivered in their barren compound, from the Indian, French and English languages.

It is with feelings of regret and justice to the ever-to-be-lamented dead, I say, that for the Geological and Mineralogical information herein presented in a condensed form, I am indebted to Dr. Houghton's Official Reports, and perhaps, in a strict construction of those terms, am, in other parts, more a compiler than the

THE AUTHOR.

Reprinted 1976 by
BLACK LETTER-PRESS - Grand Rapids, Michigan
L.O.C. 76-27042
S.B.N. 0-912382-20-1

Cover: Robert Nelson

CONTENTS.

PART FIRST.

CHAPTER I.

Having visited the mining region the past season; having neglected no means in my power to ascertain the truth or falsity of the statements, and of my suspicions as to that, by many imagined, "El dorado" of the North; having taken great pains and labour to arrive at local facts, under the recollection of the "many grains of allowance" with which I had received the best authenticated statements myself; having no interest of name or nature in the effect my statements may produce, except that in common with all other citizens, in an early and prompt development of those internal resources of our country which will leave us independent of foreign supplies; I comply with the wishes of many acquaintances, who desire the information I have obtained, too lengthy in detail for the frequent repetition necessary to gratify personal friends only, by this publication of facts as I found them, without "fear or favour," "nothing extenuating, or aught set down in malice."

I am impelled to this publication, too, not more that the truth and real merits of the country may be known, than to show, by such deductions as the careful reader will make, that "all is not gold that glitters." The *realities* are sufficient ("and to this end must we come at last,") to insure all that should be desirable to accomplish, except the ulterior objects of those whose "baseless fabrics of the brain" must fall, and visionary schemes, in the end, "come home to roost."

1*

Upon the capitalist is the dependence for the necessary means of developing our mineral resources, and upon correct information only can he be expected to furnish it. Overwrought and gilded statements, through whose glossing the careful and discreet man sees their hollowness, may deceive many, and accomplish the ends of some, but will also retard or prevent great results, which candour and truth might have produced, but for the doubts which discovered duplicity throw in the way of investigation.

Congregated capital has been necessary in all countries and times for the development of mineral resources. Nature, in her organization of matter, decreed it "when the waters covered the face of the earth;" and, as much as she has favoured this region over all others with her mineral wealth, there is also, as elsewhere, written upon it, "Thou shalt eat thy bread by the sweat of thy brow." Abundant and pure as are our lead and iron, our copper is not less so; and if there is one fact which characterizes the bounty of nature to ours over the mineral of *all* other countries, it is that *fact and peculiarity* of our Lake Superior native copper, that it is in *no instance contaminated with alloys of other metals.* The assertion of which fact, when made by Dr. Houghton, was treated as a burlesque by scientific men at home and abroad, who called it "backwoods mineralogy."

His representations as to the great abundance of copper indicated by "surface appearances," were treated as "new country stories," and Dr. Houghton, smarting under this ridicule, pursued his researches for ten successive years before his reports elicited any public attention. He has gone down to his grave in those depths, though immeasurable, and upon a rock, though unseen, which he knew and could determine in his system of philosophy, as well as if "the waters rolled back" when he came to their margin. He has gone, too, in the day when that future he had so long and confidently anticipated—was come, which, by its developments, was about to consummate the silent but prevailing ambition of fifteen years of toil, leaving one point only fully established—that the accepted systems of geology and mineralogy are in many particulars inapplicable to the scene of his labours—of which the above is one proof.

True it is, and lamentable too, that wild and exaggerated,

not to say entirely false statements have been made of mineral wealth there, to be heaped together without labour or means, which could not fail suddenly to enrich the fortunate holder of a few shares of some particular stock.

Between the extremes the capitalist must designate, or nothing can be done in bringing out the wealth of such mines as do exist and have a value. Many have already embarked, but nothing like a number adequate to the field presented. Extreme caution is not a fault, but often loses what investigation and promptness would have garnered.

The first workers of our iron almost invariably failed for the want of means, and from an influence felt by them from afar, but never seen—neither of which exist in reality now. A few years will, as in lead and iron they have, show our capacity to supply the world with copper, not less imagined at present, than it was that we would now do it with lead, when the law of 1803 was passed, with a view to ascertain if we could not produce a portion, however small, of our own consumption. Look now at the mighty result. We send it to China, and above it is piled our cotton cloths and wooden clocks, and who can say that copper will not follow?

Lead is worked, and many fortunes have been made in its production. Copper must be worked in the same manner. The mining is the same—the smelting differs—but copper ore is worth vastly more than lead, while the cost of mining is about the same. This reduces the whole subject to one or two questions, viz.: Is there copper ore as stated? and, where are its locations? These questions the reader will find answered in the following pages; and if he will read them carefully, and qualify himself, as he may, by their contents, he will be able to determine correctly between propositions now or hereafter made for investment, which are real, or which are "kiting." If he is too much engaged to do this, but goes in, hap-hazard, he may hand down at his death some "shares" which had long before formed fellowship with North Carolina "gold stock."

I am not about to advance objections to, or recommendations of, any companies, other than may exist in a list of companies working, and which is not intended to be invidious, if it shall be so construed. I give the list with the object of showing what has really been done in the country; and there are other companies, who have equally favourable positions, prospects and leases, who will another season proceed vigorously in their contemplated operations.

That there are differences between the companies is matter of course; and those differences consist in some leases being for three miles square, and subsequent ones for one mile square. Some companies are located upon good harbours, enjoying facilities which others do not. Some companies have several of these three mile leases: some companies have several one mile leases. A company may have only a single permit for one mile square, which may again in some part, not forty rods square, have more value than another company's possessions covering miles of land and the best harbours. By this it will be seen how necessary it is to be well informed as to localities, geology and mineralogy, as well as facilities, which are so materially to affect the value of points proposed for the investment of capital.

That the reader may commence his voyage understandingly, before we start to coast round the shore, I have presented for his careful perusal with the maps, a broad view of the whole country, defining the location of the various strata of rocks, of which the mountains and valleys are composed, and where those are met with.

I have given the different kinds of rock, and where they occur, together with the mineralogical contents of those rocks, immediately following; so that wherever any of those rocks are found, he may look for their described contents. If those are not there, none other will be; *vice versa*, wherever described minerals are found, there look for the rocks to which they belong.

A close study of the few pages here presented upon this matter of "formation, place, and contents of the different rocks," will enable you to see clearly your way through the country, and also the contents of the hills and mountains, and bottoms of the lakes. A small fragment, picked from the beach at White Fish Point, may have been years upon the journey from the place of its detachment, and that place may be hundreds of miles from you, which you have never been nearer to than at present; still, you will be able to put your finger upon the spots of the nativity of its kind.

So will it in a very short time be with regard to specimens

of copper. A careful perusal of the contents of the different rocks, and a familiarity with specimens which I have deposited at the American Institute, will enable you to tell at a glance what rock, region, and often the vein, a specimen has come from; and what has come to be called "*grafting*," or taking specimens from one location, and pretending that they came from another, where the rock is different, you will soon detect.

I have also given a list of native metals, their appearance, colour, consistency, gravity and situation found in.

I present a list of terms used by miners, with their signification, accompanied with certain geological and mineralogical terms which occur in the pages. I also show the entire coast, the rivers, harbours, islands, Fon du Lac and St. Louis rivers.

To the traveller for pleasure, let me say a few words. When you shall have read the round upon which I have taken the coaster, you will probably shrink from the toils of following the shore, and wish to go *direct* and *quick* from place to place, or tarry a time at one place and then go to another—in either case, there will be every provision next season. A steamboat, large, staunch, commodious and safe: a propeller with all these qualities also, and a number of very convenient schooners, as will be seen by the list of vessels on Lake Superior. If you are in pursuit of pleasure, whether lady or gentleman, you can find it in the Lake Superior region, provided you can be pleased with grand scenery, water-falls, lakes and mountains. You can ramble in search of Agates and Cornelians, in which, of all I have seen engaged, I have never known one tire of the amusement yet; to become so fatigued as to stretch upon the pebbled shore and search within the reach, then crawl a space and there search on, and still as anxious and intent as when first beginning, till time, who is flying while you are absorbed and unconscious of his flight, begins to dim your vision with a declining sun, and weaken discrimination of the prize you seek;—then rousing to consciousness, you see the sun that hung high in the heavens when you commenced your search, just sinking in the waves, and reflecting, you seem to have been away in another world and just returned; you look about for some known object almost doubting your identity;

reluctantly you shape your course for home, but hope lingering hangs upon the way; though fatigued and

O'erloaded with the selections fancy found,
You'll pick, look—'one more,' you'll say—
"Another with those 't are to be ground,"
Or examined dry, be thrown away.

This is Agate hunting, as all will testify who have tried it, the most fascinating and bewildering yet certainly innocent amusement. I have seen a staid and dignified old Governor stretched at length upon the shore from very exhaustion, absorbed and lost to every thing but examining agates, consuming half an hour in scrutinizing and admiring the variegated tints, the beautiful blending of shades and colours, and the regularity of the myriads of diverging and concentrating lines of different colours in an agate he gazed upon, not larger in circumference than a dime; or tired of this you can wander away with "hook and line," to the bright and beautiful lakes that lie among the hills; or take your gun, for

The Pigeon and the Partridge's there,
The wild Duck and the timid Hare—

but no snakes! I have never heard of any in the country. Or take a bark canoe, which two or three trials will make you at home in, for they are much easier to get the "hang" of, than most persons suppose; go to the adjacent islands, run into the caverns and grottos which cannot be reached in any other way. You may find rare agates there after a gale, and when you return, keep along the shore and examine the bottom marked by the white spar veins discernible at thirty or forty feet deep, or nearer shore with a forked stick bring up the stones you fancy agates beyond the reach of those on shore, and when you get back you will have an appetite; the tonic air of that region, and the water, will make a new being of you in a few weeks. The air is bracing yet soft, and is pleasant in "dog days," without producing that faintness and lassitude of the warm weather you have been used to: and the water—well, you will not be singular, you will then say you have never drank any water before; and when you return whence you came, and again drink of that you once thought delicious, you will condemn it as an adulteration, or spurious.

To the invalid, I have a few words to say, for *his* information; I am not "cracking up the country," for I shall write nothing that all who go, will not find as I represent it, or all who have been will not confirm, either on this or any other subject of their acquaintance which I treat upon. To you I say, go then; although your health is impaired you cannot be injured, and I know one gentleman who had been south; had been to Havannah without any benefit; one season on Lake Superior restored him, as he said, to comparative health; he was a companion 'du voyage.' I don't know why it should not relieve consumptives as well as others, all who go there declare *they* feel much better—and I *know* I did. I was under a slight chronic and cutaneous affliction; I was told that the lake Fanny Hoe, on which Fort Wilkins stands, was so much impregnated with mineral, that a soldier died from drinking of it. I asked if the other soldiers had not drank it as well as he, and was informed they had, but its water had since been analyzed and found to contain too much copper. Upon this I resolved to drink the water of Fanny Hoe lake, brackish as it might be, and continued to do so while there, and I firmly believe it to be beneficial in cutaneous affections. I didn't die, as the soldier did, but felt better every day I remained. One thing is certain—the half-breeds and natives all live to a great age, notwithstanding their exposures: and sickness from fevers, colds, inflammations and agues, is scarcely known; the healthy and ruddy appearance of all you meet will be a stronger guarantee and more satisfactory evidence, on your arrival, than any philosophical reasons I am able to give. I am assured there will be prepared early next spring, accommodations for travellers and sojourners, at all the places desirable to stop at. Even now, every hospitality is afforded which can be, but especial preparations will be made early, for next season's "passengers and baggage to and from the steamboats."

The maps herewith presented, are the most correct possible to be offered at this time. The chart of Lake Superior is perfect. It may not be generally known, that Lake Huron and Lake Superior have been as perfectly surveyed as science and time can do it, and this chart is a result. Lieut. Bayfield was several years engaged by direction of the British government upon this work. The chart herewith, is a copy reduced by the "London Society for the diffusion of useful knowledge," from the map reported by Bayfield as a result of his labours. Its accuracy is attested by all of any acquaintance with that Lake and shores.

The map of the Mineral Region and Locations made, was taken from the standard map in the office of the Mineral Agency at Copper Harbour, containing the locations made up to the 1st day of November last. I have added to it many lakes, harbours, rivers and islands. So far as leases are granted, this map is perfect in designating locations, because the survey is made by a United States surveyor before the lease issues, and their locations may be depended upon.

The numbers and squares designating Locations, are in the main likely to be correct when in the vicinity of Leases, for in such cases the locators have established corners and lines of those surveyed Leases to work from, in measuring and describing their own. But as I am dealing in facts with the reader, I must say, what experience has already shown, and the future must confirm, that there are difficulties in designating upon a map, without any great right lines fixed to govern it, the precise position of an isolated survey. The location may be correct in all particulars as described, the starting and governing points all known, and the measurements correct, while its position upon the map would be erroneous from the want of a base and right line whose measurements would fix it accurately. Another, and which may hereafter be a prolific source of difficulty and contention, when positions come to be defined as described, by correct surveys, is the variation of the magnetic needle in the mineral region; accompanied too with the known difficulty of measuring distances correctly in the thick bushes, except by the most patient and careful attention to accuracy, obtained slowly and by great labour, and an adherence to an undeviating system; then seldom, or never without the solar compass to determine at every station the "variation."

In this particular, of such paramount importance in making a location, some I know, and many I doubt not, have been negligent—defining their starting points, courses and distances by a *pocket compass*, and *pacing* their measurements. All can judge of the probable results in such cases from a statement given me by Mr. Hill, one of Dr. Houghton's head surveyors, that there were places in the mineral region, where the compass would, following a given course by it, lead on a semi-circle in running one mile. Now suppose an individual, in locating a permit with an ordinary surveyor's compass, has started his course east at one of these places, and has run a semi-circle; he does not know that he has so ran, and consequently says in his description, "from thence, east one mile to a blazed Spruce tree, which is the north-east corner." He does not know but it is so, and goes and marks his position and gets his permit certified with this description. When this location is run out by his given courses and distances, with a corrected compass, his "blazed tree" is found to be about three-fourths of a mile from where he supposed it was; and the mineral he intended to locate upon, is off of his location. To what extent this is the case I have no means or disposition for estimating. Again, there have been persons lingering about the office at Copper Harbour, with pockets full of Permits for persons never in the country, and never intending to be there, who came to be known by the name of "*Pawnees*," from their putting their paws upon the shoulder, and "a word in your ear" to every explorer who had really made examinations, the moment he arrived. One would not be done with his "one side" inquiries, before another *paw*, and, "a word in private"—each hoping to get an unguarded word on which to locate some of his "friends' permits." I was told, with what truth I know not, though my authority was an 'old 'un,' that some of the Pawnees have been "awfully stuck," to use his expressions; "for," said he, "some of their permits are located *where they can't touch land up, nor down, nor sideways—and others ain't nowhere.*" This might have been true in some particulars, but usually the "Pawnees," where a location was marked on the map by an explorer, located around that, and bounded from its description in making out their own, which I have no doubt in many cases will cover the spot the explorer intended. An anecdote is related of two of these Pawnees in whom the habits of "privacy" was extreme, meeting when each had communications for the other. They met in an open space, twenty rods from any person, and after whispering in each other's ears as if announcing their mutual desire to be "private," locked arms and walked to a more remote position.

CHAPTER II.

A VIEW OF THE SHORES, RIVERS, BAYS, HARBOURS, AND ISLANDS OF LAKE SUPERIOR.

ALL explorers, with scarcely an exception, when they set out for the first time, encumber themselves with about as much that is useless as that is useful. I will therefore state what is *necessary* only, as the necessaries are usually found quite burdensome enough.

If you intend to "coast round," you must provide at the Sault St. Mary a bark canoe sufficient for your number, and provisions for the estimated time, which must allow for detentions, &c., until you reach Grand Island, as there, at Granite Point, (Dead River,) and L'Anse, provisions will probably be obtainable next season. You will require a tent, two blankets for each, a camp kettle, frying pan, tin cups and plates—if you are nice, a coffee-pot, ground coffee, and sugar; but the plates, coffee-pot, &c., may be dispensed with, and often are, the kettle or frying-pan being the dish from which each helps himself,—the knife is in your belt. The sail of your canoe is laid upon the boughs, which are first spread upon the ground in the tent, and then your blankets. At landing, the canoe is not allowed to touch the bottom, but you get out into the water and unload it, which is then lifted out upon the shore, turned bottom upward, and your stores are secured under it; your tent is pitched, a fire built, &c. This is all, however, the work of the voyageurs, who, from practice, will despatch it with a facility that will quite astonish a new traveller.

Your clothing should be a pair of thick-soled boots of cowhide; no stockings are required, but most persons wear them, and consequently have the nightly recurring duty of drying them *almost* dry, and thus putting them on damp in the morning, producing a contest between the boots and feet of entrance and resistance, rather disagreeable. A pair of pants of cotton canvass, and a coat of the same to reach below the knee, with side and breast pockets. Cotton canvass is found to be as good as anything to turn water, and the best to turn the brush, which is a work of labour and perseverance often to be endured for hours in exploring. A red flannel shirt, will not require so often washing as a cotton one, and "it is always warm and dry, though never so wet and cold;" besides, washer-women are rather scarce, and when you have performed the office yourself a few times, you will become less fastidious in relation to such matters. A red woollen comforter. No suspenders; they will confine you in crawling under logs and limbs, and through difficult passages in the cedar thickets. A belt, carrying a hatchet and knife, buckled round the waist, will sustain your pants, and allow free exercise of your body and limbs. A wool hat, with wide brim and low, round crown, is the best to turn rain and brush. A pocket compass, and perhaps a pipe, completes your equipment, saving a few fish hooks and line. These are the really necessaries, though most travellers are not content with them. Experience, however, shows the necessity of being divested of everything which may be dispensed with; for, portages and journeys have to be made, in which every thing, even the canoe, must be carried for considerable distances, on which occasions "blessed be nothing." Habit, however, brings power of endurance which many would not believe, and I have seen a packer, himself weighing less than 145 lbs., who could take upon his back 200 lbs. weight, and make good time upon the portages.

In coasting, it is necessary to have at least one good "*voyageur*," as they term themselves, who will most probably be a Frenchman or a half-breed, who understands the coast and weather-signs,—superintends the unlading and camping—interprets—knows where fish may be taken—the proper places for landing, and whose counsels—as to whatever implicates safety or convenience must be followed, and with rare exceptions may be; for, experience makes them wise in things which gives their knowledge the appearance of intuition, when contrasted with their general endowments—as, the coming of a storm, the probabilities of reaching a particular necessary landing-place by a given time, &c., &c.

So much depends upon the voyageur. If the interior is to be visited, he can generally tell you the route, nature of the country, &c. It will sometimes occur that you will be detained for two or three days at a stopping-place by high winds, but this is a rare occurrence. At other times, when the weather is favourable, you will sail on during the night, which is determined by the voyageur's opinion of the weather, and knowledge of the coast to be passed.

Being in all things prepared, you set out from the Sault St. Mary, generally in the afternoon: for starting is rather difficult, from the fact, that your voyageur will be drunk from the time he is engaged until your departure; and you will start with the purpose of going a few miles that day, and having him sober next morning, when the voyage really commences.

You will be surprised in the morning, to see what a change one night has made in him. Until now, he has taken no interest in the preparations for the voyage, but here, he is aware that his responsibilities have commenced, and, instead of the beastly and besotted creature he was yesterday, he is taking a mental inventory of whatever preparation has been made; each article of use is minutely examined; alterations and improvements, if necessary, are made; occasionally he casts a look upon the circle of the heavens for indications, as a book long studied; you "turn out," and are surprised that he has been up with the dawn, and breakfast is about ready;—this over, if the weather to his mind is unpropitious, or he has private reasons, he will find that the canoe requires some tightening, and, putting his lips to the suspected spots, will find, or pretend to find air-holes. These he stops by taking a burning brand and holding it near the gummed joint, or seam, and when the gum is softened, he rubs it down with his wet finger; if this do not detain him until his *mind is fixed*, as to the weather and other matters, he will find other expedients, and without knowing *his* reasons for delaying, you change mind in regard to him, and believe him a lazy fellow;—but you can't go back, and must endure it. While you are arriving at this erroneous opinion of the Indian, he is doing exactly the reverse with you, and his advantages in that contest are far greater than yours. This is your first voyage and first acquaintance with his kind, but *he* has been all his life in this business, and thus made long voyages in company of some of the most qualified and enlightened gentlemen of the world;—perhaps when a boy he was with Porter running the boundary; later, he might have been a man with Bayfield, and every year has thrown him with men to draw companions by, precisely what he is now doing with you. Now mark the necessity of his showing *you* well;—you are under *his charge* in fact, for this business is his living, and his reputation as a *voyageur* has procured his employment for you. You did not hire him—Mr. Livingston, the American Fur Company's Agent at St. Mary, probably furnished him, and if he did, he knows him better than you ever will, and knows you are *safe* with him. But why does this Indian wish to know you? may be asked; he merely wants to know your calibre, if I may so speak, and then he fixes you at your proper place on his scale of men, in order to know when and on what points to consult you,—to know what dependence can be placed upon you in an emergency; and these points, you may be sure, are all settled in his mind before he passes White Fish Point. He now knows you by contrast; you only know him as you first saw him, a besotted, helpless being, and you cannot help divining him as such. You may go the whole voyage, even repeat it, without an occasion presenting itself for that Indian to exhibit what he really is; but if one should come, in which he is required to put forth his powers of judgment, skill, or endurance, to preserve the safety of one committed to his charge,—then, and only then, can the true character,—the endurance and self-devotion of those hardy and honest hearted *voyageurs* be truly known. Gay and mirthful by nature and habit,—patient and enduring at labor,—seeking neither care nor wealth, and though fond of their families, "take no heed for the morrow." Such are the "*voyageurs*" of the Lakes, a distinct and different people from all others upon the globe.

These *voyageurs*, or *Coureurs des Bois*, as some call them, have many sketches and songs, to the chorus of which they ply their oars, and cheer their toil. The following two are much in vogue, and heard from a full crew in a still night, well might have given origin in the heart of Moore, to his "Canadian Boat Song." These were formerly published in a work entitled ‘ Tales of the Northwest," but are still in use,—one leading with the text, the others joining in the chorus. Dr. Houghton's crew woke me at the dawn of day with their song, the last time he departed from Copper Harbor.

1.

Tous les printemps
Tant de nouvelle.
Tous les amants
Changent de maitresses—
Le bon vin me endort
L'amour me revielle.

Tous les amants
Changent de maitresses,
Qu'ils changent qui voudront
Pour moi je garde la mienne—
Le bon vin me endort
L'amour me revielle.

2.

Dans mon chemin j'ai rencontre
Trois cavaliers bon montees
Lon, lon laridon daine,
Lon, lon laridon dai.

Trois cavaliers bien montees
L'un a cheval a l'autre a pied
Lon, lon laridon daine
Lon, lon laridon dai.

CHAPTER III.

The *Sault St. Mary's* was last year reached by taking steamboats of the Chicago line at any of the ports, and from which passengers landed at Mackinaw, which the tourist finds a most beautiful and pleasant place in hot weather, which was the resort and sojourn of many last season. From Mackinaw a steamboat, I am assured, will run *daily*, instead of tri-weekly, next year, to the St. Mary's. This route, for beauty, boldness, and diversity of scenery, is unequalled in the world as a panorama. It is for most of the way among the "*Ten Thousand Islands*," as reported by Lieut. Bayfield of the Royal Topographical Engineers, who was for several years employed by the British government in surveying Lakes Huron and Superior. He reports having been upon *twenty thousand islands* in Lake Huron, and that ten thousand of them lie in and around the embochure of River St. Mary's into Lake Huron. The steamboat is continually winding its devious way among these islands of various sizes, standing amid waters of vast depth at their very margins. Some are large, and covered with forests of sugar maple; others bring forth only the cedar from the opening crevices in their sides; others bear upon their brows the stinted pine and silver fir; while others are but stupendous piles of naked rocks, whose perpendicular sides, as you pass them, sometimes within a few feet, show by their disjointed seams a "crash of matter and a wreck of worlds" must have raised them in their giddy heights and overhanging positions. Anon, the "slow bell" rings, and gradually the boat turns at a right angle from a lake-like space, and enters a labyrinth of islands, from out of which an inexperienced pilot might not find his way for days. On one hand open deep caverns in island rocks, and on the other are spread out in solitude the small grassy meadows where foot of beast never trod. Along Lake George, upon the projecting points and island shores, are often seen the Indian's wigwam. There these children of nature ply their only toil, procuring food, mainly from fishing, content in primitive simplicity, and free from the great sources of unhappiness and care to the civilized—*fashion*, and the accompanying responsibilities.

St. Mary's has been the location of a few Indians from the time of the first travellers, and the resort for large numbers of the natives for fishing at the foot of the rapids. In the sale of the lands to the United States a strip bordering the north side of the Sault was reserved as a camping and fishing ground, which is known and regarded as the "Indian reservation." There is also another reservation of twenty square miles, including this first, and the present village of St. Mary, Fort Brady, &c., known as the "government reservation." The Indian reservation is included in and bounded by the principal street or road over the portage, on the south, and the Sault or rapids on the north, and by their head and foot, including an island in the stream, running nearly the whole length. This reservation is not interfered with by the whites, but the half-breeds are beginning to erect buildings upon it. It is upon this reservation the proposed St. Mary's Canal must be constructed when done Upon the United States, or government reservation, are built the forty or fifty houses which constitute St. Mary's, containing a population of 200 inhabitants of all nations, colours, grades, and languages, exclusive of the Indian lodges. Their titles are those of "squatters," and are conveyed by quit-claim of pre-emption, which are recognised, however, in all questions of possession by the local courts, and quite a speculation is carried on in "corner and water lots." In 1721, Charlevoix visited this place, said to be a short time after it was founded. In 1762, it contained but four buildings, two of which were occupied as barracks, with a stockade fort and garrison. That old fort is entirely destroyed. In 1820, there were six or seven dwelling houses. It is in Chippeway County, Mich. There is also a mission school under the charge of Mr. Brookway. The American Fur Company have a station here under the management of Mr. Livingston, a business man, and gentleman of long residence and experience in the north-western country. The United States have also a mineral agency, filled by Mr. S. M'Knight, and a post-office, custom-house, and Indian agency. There are several forwarding establishments, two public houses, (the poorest house is the best tavern,) several stores, &c.

St. Mary's is accessible to the British from the north, it being but ten days' travel by boats between James' Bay and this place, from whence the supplies of the Hudson's Bay station at Pentenguisheen, on Lake Huron, are received, and is the route for all of their people to England, from the northern and upper stations. Through here, too, could an army be thrown, by a little preparation, upon our northern frontier, before their landing upon this continent was known at Washington.

From St. Mary's the river is wide and beautiful to Point au Pins, (Piny Point) where we left the canoe and voyageur, and from there we will now proceed on our voyage, and nine miles brings to Iroquois Point, opposite to which, on the Canada shore, three miles distant, is Gross Cape, (Great

Cape.) The event which gave name to Iroquois (Nado-wa-ga-quin-ing—the place of Iroquois bones,) is thus described by H. R. Schoolcraft, in a discourse before the Historical Society of Michigan, and describing the victorious and bloody advance of a large Iroquois war party (in 1680) to this place, —he says:

"Their passage through the river, and the audacious and reckless spirit which they had everywhere manifested, had been watched. The Chippewas hastily mustered their forces, and prepared to follow them. When they had reached the head of the straits opposite the Iroquois camp, the weather became threatening, and it was debated whether they should not defer their passage till next day. In this dilemma their prophet, or seer, was appealed to, and he, after the usual ceremonies, declared a favourable omen. They awaited the approach of night, and embarked in two divisions. The darkness of the night was extremely favourable to their enterprise. The parties landed at separate places, and formed a junction in the woods in the rear of the Iroquois camp. The prophet here declared another favourable omen. They then sent forward some scouts to observe the condition of the enemy, who appeared totally unconscious of danger, and were still singing their war songs. It was determined to remain in their concealed condition until the enemy had gone to sleep. It then commenced raining. They advanced in the rain and darkness, cautiously feeling their way, to the edge of the woods. They then made their onset. The struggle was fierce, but of short duration. As had been concerted, each lodge was surrounded at the same moment; the poles lifted, and the tent precipitated upon the sleepers, who were despatched as they started up bewildered and entangled in their tents. A great slaughter ensued. Very few of the Iroquois escaped to carry the news of the disaster, nor did their nation ever renew their inroad." From Iroquois Point, N. W.

White Fish Point is thirty miles, and by coast to Tonquamenon River is fifteen miles, thence to Sheldrake River nine miles, and thence to White Fish Point, being thirty miles direct, and forty miles by coast. Here is the entrance upon Lake Superior; the coast runs west with sand hills for eighteen miles, when it falls off south-west six miles, being twenty-four miles to

Two Heart River. From this the coast again runs west, and the high sand hills continue for twenty-one miles to

East Grand Marias, or Hurricane River and Bay. This is a large bay, with sandy shores, shaped very like Copper Harbour, but a narrow entrance, varying at times from two to five feet upon the bar. Upon the main shore, which continues west, are sand hills, continuing nine miles to Grand Sable, which are 100 to 300 feet high, and when surmounted at a distance of three miles back, the country is level, sandy and barren. A small lake of good fresh water is found in this plane, containing abundance of fish, which has neither inlet nor outlet. From here the coast runs south-west twelve miles to the Pictured Rocks.

CHAPTER IV.

Pictured Rocks, (La Partaille.) The best description I have met with, and which is sustained by all others, of these rocks, is the following one by A. B. Gray, Esq., one of the assistant mineral agent. He says: "Commencing a little west of Miner's River, and extending easterly for ten or twelve miles, is a high and perpendicular wall of sand stone, rising to a height of three hundred feet, of horizontal strata, several feet in thickness, coloured with various bright and beautiful tints of vegetable and mineral matters, and forming one of the most picturesque and deeply interesting natural curiosities in America. The water near its base is a clear emerald green of great depth, allowing vessels to approach within a few feet of the narrow pebbly beach here and there to be met with, and elsewhere the rock itself rising immediately out of the lake. Successive curves of half a mile in length, caused by the wearing away the soft sand rock by the waves, appearing like painted walls of an amphitheatre, and continue for nearly the whole distance, occasionally interrupted by a small stream, or cascade leaping from the precipice. When near its base in a small boat, the projecting summit of this massive structure presents a grand and awful sight. Rotundas, caves and domes with arched entrances, curiously and beautifully formed, are numerous. One cavern, which we sailed into with our boat, had an arched way of fifty feet in height and thirty in width, which suddenly expanded into a high and singularly constructed rotunda of two hundred feet in diameter. The 'Doric Rock,' the 'Pulpit,' and 'Des Partaille,' and other features of this portion of the southern coast of Lake Superior, called the 'Pictured Rocks,' altogether constituting scenery of grandeur and beauty unsurpassed. The tops of these rocks are covered with a small but symmetrical growth of silver-fir, spruce, maple and birch, and in this month (October) their rich and variegated foliage presented a strikingly beautiful appearance." Six miles is

Miner's River and Dorick Rock.—This river is very rapid near its mouth, making a quick descent to the Lake, through the Sand rock. Pursuing the coast, which is rocky, and runs S. W. twelve miles, forming the east side of Grand Island Bay, is the bottom of that bay, where the shore turns almost due north, and a small and nameless stream enters. From this north three miles is

Grand Island, which is nearly about a mile from the main land, three miles long N. and S., and one and a half E. and W., with two small islands on the west side. This island and bay presents one of the best harbors on the whole chain of Lakes, easy of access, and safe from any wind. The land upon the island is 500 feet high, and a vessel in its southern nook is almost land locked. It is fertile, and the residence of a Mr. Williams, who has considerable stock. From here the shore curves first to the southward, forming a bay, where a small stream enters, then a semicircle northward, falling quickly south, forming another bay, in which, at a distance of nine miles from Grand Island, enters the

River au Trains.—Three miles west of which, and on the west side of this bay, is Train Island, which is a small island, near the shore. The shore from this is irregular, bearing first to the north, then west for six miles, to

Laughing Fish River—Which enters a small bay, from which the coast, on which grows pine, spruce and hemlock, with a large swamp back, runs west fifteen miles, passing Little Sauble, to

Chocolate River.—The mouth of this river, which is a considerable stream, and takes its name from the colour of its water, is 146 miles due west from St. Mary's. This stream is for several miles, from eighty to one hundred feet wide, and a good depth inside; but the sands at the mouth make it difficult of entrance, even for large batteau. It has its rise in a lake at the foot of the mountains, which separate the streams falling into Green Bay, and those falling into Lake Superior, and winds its way to the Lake along the base of rock hills, which extend to S. W., forming the commencement of the mineral region, while the east side is level country. The timber of this river consists of maples, ash, oak, bass wood, and pine. The soil is said to be good. This river is the eastern boundary of, and from it west is the mineral region. From here, the general bearing of the coast is N. W., and distant six miles is one of the few rivers of the South shore which have their rise beyond the first range of hills; and here commence the Trap range of hills, rising up through the sand stone strata, which run S. W. nearly unbroken, to the head of Lake Superior.

CHAPTER V.

Dead River enters the bay formed by the south side of Granite Point and the main land, and is a good harbor, which was formerly called Presque Isle Bay, but which name has been transported six miles west, where a Presque Isle Bay is found to match it, just under what was formerly called Granite Point. Bayfield removed Granite Point six miles east to Dead river, and Presque Isle Bay; and I know no reason why Presque Isle cannot be spared in exchange, particularly as it had three names—remembering too, that all the names out of the genealogies of the Israelites, have been "used up" to form mining companies. Hereafter, let it be borne in mind that Granite Point is removed to Dead River, and Presque Isle is gone six miles west of it.

This Dead river is by no means the sluggish river its name would indicate, for it is one of the most rapid of the South shores, with cascades and water-falls. The Indian name was "Ne-kom-e-non," "the River of Deaths,"—undoubtedly the scene of several deaths of some former time, which the early French very correctly called "Riviere des Morts." Farmer's map has it "Dead River," and Bayfield has done the same, which is farther from the original in meaning, than "Living River" would have been. The harbour is known as Granite Point, upon Bayfield's chart, and as such I shall call it under the head of "Courses and Distances" in another place. This point is covered by lease, twenty belonging to the "New-York and Lake Superior Mining Company," and will be found under the head of "Companies Working." The river mouth is under the south east side of the Peninsular; a sand bar prevents an entrance, though inside for some distance it is broad and deep. The peninsular runs north and south, and the south bay is protected from all but a strong easterly wind. The north bay is filled with islands and rocks, but a good harbour may be found there from easterly weather when it would not do to enter the south bay. Eighty rods from the north point of this peninsular, are two rocks high above the water which is five fathoms, and bold, as well by the south side of these rocks, as by the peninsular, alongside of whose bank a vessel may be unladen with a plank; the course between these rocks and peninsular is N. W. and S. E.: the schooner Swallow, was beat through here at 12 o'clock at night, being the 27th October, wind light from S. E.;—a shoal, or reef, puts off east and north from the easterly one of these rocks, a distance of a quarter of a mile, but N. E. from south bay, will be sufficiently east to clear it. There is also shoal water but sand bottom putting off east from the south point of the peninsular; but from the mouth of the river an easting of 80 to 120 rods is required, or to five fathoms when it will do to come up N. E. by N., which will carry you clear of every thing and give a good offering at six or eight miles to shape your course. No soundings will be found in entering from the N. E. until you pass the shoal or reef from the eastern rock, but you run for the centre of the bottom of the bay until you get bottom which will probably be five fathoms;

opposite the mouth of the river, the water shoals gradually to the shore unless you get too near the sand shoal from the south point, already mentioned; there are several islands near the shore above Granite Point, and one considerable one about one mile due west from the north extremity of Granite Point, between which and the main, of easy entrance on either side of the island, is a good shelter called Talcott Harbour, beyond which, from Granite Point, three miles, is what is now known as

Presque Isle.—This is covered by lease, twenty-one belonging also to the New-York and Lake Superior Mining Company, but they are not working there. About seven miles from the shore of Presque Isle east, and ten miles due north from the northern extremity of Granite Point, is *Granite Island:* a pyramid of rocks rising about one hundred feet with a few trees upon its east end, which contains about two acres; the water is deep around it, and though dangerous in dark nights and fogs, may be seen for a great distance; the trees upon its top discovered above the fog, was the first evidence of land we saw for two days, having, however, discovered our whereabouts that morning, to be Huron Island by the echo of our gun, and were running for Granite Point with a good 'look-out.'—From Presque Isle it is nine miles to

Garlic River, which is an inconsiderable stream. The coast is of red sand rocks and granite, with small islands, and from Granite Point runs N. W. for fifteen miles from Presque Isle to

St. John's River, which is represented by Bayfield as running a considerable distance into the country; but little is known of it except that it is rapid in its approach to the Lake, from among hills, which rise from six hundred to nine hundred feet above the Lake. From St. John river, the coast is irregular, passing north round a peninsular, and round a considerable bay to the south, and then irregularly west: fifteen miles thence to

Salmon Trout River, an inconsiderable stream, coming from a small lake in the first range of hills, which here come close to the lake with a height of 300 feet. Six miles is

Burnt Pine River, also an inconsiderable stream. Nine miles farther is

Huron River. It is another of the few rivers of the South shore which rise beyond the first range of hills. It is a considerable stream, rising in the second range of mountains, and running more than 100 miles in its course to the lake. Its source is in the same ridge of those of the Ontonagon and Minomenie, the first falling into Lake Superior 100 miles west from the Huron, and the latter discharging its waters into Green Bay. The Huron is navigable for canoes for a considerable distance, and is said by the voyageurs and explorers to pass through a country of good soil and timber, abounding in game, and which is part of the northernmost hunting grounds of the Minomenies. Some mineral locations have been made, and a settlement commenced at its mouth. Two miles from the main land, N. E. from the mouth of Huron River, are the

Huron Islands, on the south side of which is a good harbour, of easy access from the west, and also from the east, by giving the east point, from which rocks and shoals put out, a good wide berth, which there is no difficulty in doing. These islands are high, as is the main land on which are those hills called the

Huron Mountains. These are the highest lands of this section, being 400 to 800 feet. From Huron River the coast falls off south-west, then south for twenty miles, forming a deep and narrow bay, across whose mouth from Huron River six miles is

Point Abbaya, which is the west shore of the bay just mentioned, and the east cape of Keweenaw bay. The shore of this cape runs near south-west forty miles to L'Anss, with the exception of a peninsular of two miles in length, about half way down the bay, by some called the Traverse, as the cape east is usually followed to this place, from whence the Keweenaw bay is crossed, making from Point Abbaya twenty-one miles to

Portage River. The mouth of this river is narrow and sandy, and there are four feet water at its entrance of Keweenaw Bay. It leads into

Portage Lake, running N.W. and S.E., which has several small streams coming into it from Keweenaw Point on the north, and others fróm the main land on the south, through which, twenty-one miles, and across a portage of one mile, is reached Lake Superior, at a point sixty miles S.W. from Copper Harbour.

This is the usual route of the voyageurs to Fon du Lac. We will now go down to L'Anse, and from thence coast up around Keweenaw Point, till we come to the Portage again.

CHAPTER VI.

At the bottom L'Anse Bay, which becomes quite narrow, enters the Keweenaw River, as it is called, a small stream, near the mouth of which is a Fur Company's station. Five miles from the entrance of this stream, on the east side of the bay, is the Methodist mission, under the charge of Mr. Brown. On the opposite side of the bay is the Catholic school, under the charge of Mr. Barrigaw, late of Philadelphia, who has erected the buildings and supports the establishment from his private fortune; he is universally represented to be a finished scholar, a perfect gentleman and true philanthropist, whose labours, examples and teachings, have gained for him the confidence and deepest regard of the natives, and highest esteem of all so fortunate as to make his acquaintance. At L'Anse is an Indian agency, a blacksmith, and a farmer to instruct the Indians in agriculture. The Indians already bring the finest potatos to market at Copper Harbour, eighty miles in their canoes. From here the west coast of the bay runs due north to Portage River.

The country north of this river and lake is Keweenaw Point, 60 miles N. E. and S. W., and 30 miles across, which has been constituted Houghton County. Among the small streams entering Portage Lake from the main land on the south, are the Portage and Sturgeon, and others from the peninsula on the north.

Following up the east coast of Keweenaw Point, 20 miles above Portage River, are found two small islands; and eight miles further on the shore, a peninsula running a mile or two into the bay; and 18 miles further, what is called the Cascades, a bold sand-stone shore four hundred feet high, extending five miles, which the voyageurs select a propitious time for passing in their canoes, as there is no landing in a storm; however, few or no accidents have ever occurred, for in addition to caution in starting, generally every nerve is plied, and the bark canoe glides past the frowning battlements in half or three-fourths of an hour.

About 20 miles south from Copper Harbour, round by the coast, and six miles across the point, is the mouth of Little Montreal River, having a fall of 40 feet perpendicular, and entering a very good harbor from north and westerly weather, in which Copper Harbour cannot be entered safely, called Bay Bris, or Rolling Pin Bay, named from its being the shape of a rolling pin.

Four miles further is a bay formed by a small point, which affords a good resort and shelter from all winds west of north.

Manitou Island, (or Great Spirit Island,) is about eight miles circumference, three miles east from the extremity of Keweenaw Point. A small island or rock, with a reef running south for half a mile, is on its west end next the point, just east of which, however, near the Manitou, is a tolerable shelter from a northeaster.

The Stanord Rock.—This is a granite rock, 60 or 80 feet long, ten feet wide, and four feet high above water. A shoal or reef puts off to the N. W. for 80 rods, but on the other sides the water is deep. This rock was discovered by Capt. Charles C. Stanord, at 4 P. M., August 26, 1835, then master of the brig John Jacob Astor; first discovered at a mile distant dead ahead.* Its position is 29 miles S. E. half E. from Manitou Island.

* There are three brothers of the Stanords, all captains upon Lake Superior; Charles, Benjamin, and John, who have all been for more than ten years there, and are the oldest and best pilots, careful and good seamen, and gentlemen. John, who is master of the schooner Swallow, has come to be called the "pig let out of a bag," from the fact that after running for days together in the dense fogs without land or bottom, he scarcely ever loses his "reckoning;" and if he does, he fires his gun, and by its echo he knows if he is near land—and if so, determines by its vibrations the coast he is upon, and thence how to shape his course. So too with the brothers. Long years of experience in feeling their way with the gun, enables them to judge of the land when the sound rolls along it, although they cannot see it, with an astonishing certainty. This occurred after being three days in a fog without soundings or sight of land. "All keep quiet," said Capt. John, "till we ascertain if land is near and where we are; it ought to be pretty close—fire!." The blaze opened the thick fog before the gun; the sound rolled over the water for a second, then a crash! an-

CHAPTER VII.

Copper Harbour.—Is the name by which Fort Wilkins is now known. The entrance is wide and easy with a wind from any point east of N. or S., but may be entered with care with almost any wind, if the sea is not running too high, the entrance being wide enough to beat in between a large rock at the end of the reef which runs east from Porter's Island, and the point of land on the east side of the entrance. The course is S., until within 30 or 40 rods of the *Astor Rock*, then it is west 1 1-4 mile to the anchorage at the head of the bay, where there is safety from any weather. The difficulty in entering in a heavy sea from the north or north-west, is in hauling up or coming about to beat up the bay to this anchorage; for if she misses staying, or gets "in irons," the rocks are too close, and the holding ground is not good against the sea that in a gale rolls over the reef, which was the cause of the loss of the brig Astor, which dragged her anchor and was stove to pieces upon the rocks, where her skeleton now lies, just in front of the entrance of the harbour. The harbour extends about 2 1-2 miles east and west, and 3-4 of a mile wide—a parallelogram.

"About 80 rods south of the harbour, and 15 feet above it, extending east and west, is Lake Fanny Hoe, one and a half mile long, and 60 to 80 rods wide, with hills 300 feet high falling suddenly to its south margin; while at both its east and west end several ridges meet. Upon the highest peak of this mountain the first rays of the morning sun were brightening the yellow tinge of the foliage, and magnifying every object upon the summit; the base threw its dark shadow upon the lake, that mirrored it with a truthfulness

other—another—then the sound rumbled along to the eastward. "I thought so—the Huron Islands—did you hear it rattle among the mountains, and then roll along the shore to the north-east. You can't hear it this way, for there is no land, the bay makes away to the south—ready about." Then the short explanation was interrupted by a voice, "All ready forward," and "hard a-lee," brought the Swallow into the wind. Her sails shook for a moment, when she fell off, and we commenced "long and short legs" for Granite Point, where our iron pilot's loud hail brought forth a light by the miners, about 12 o'clock at night, which enabled us to "beat through" between the peninsula and two outer rocks into the south bay, under Granite Point.

which brought the reflected mountain side to almost cheat the understanding, and tempt the foot upon the treacherous shadow. On the right are the neat white buildings of Fort Wilkins; from the staff the soft morning breeze spreads gracefully the 'Union,' displaying the stars and stripes, a guarantee that order and protection are established in this wilderness of lakes and mountains; and, at the instant, up, far above the mountain's top, in ascending circles, rose an eagle, 'patriarch of his tribe,' who seemed

> 'Journeying with the morning sun,
> Our northern bonndary line to run.'

"The solemn quiet that reigned upon the elements was deepened by contrast, when had ceased the 'morning drum-beat,' that brought forth the cleanly equipped soldier and gaily plumed and appointed officers *en parade;* and there the miners landing from the mountain side, from their subterrancan toils released, saunter with leisure steps and idle gaze; all which proclaim this a Sabbath morn—observed by rest, as it only may be,

> 'Where the sound of the church-going bell
> These valleys and rocks never heard.'"

The Garrison is under the command of Capt. Albertis. Mr. C. Brush is sutler, having a store at which may be found every thing necessary to comfort, and, I might add, luxury, and is himself a gentleman possessing great business qualifications, whose accommodating disposition and frank deportment insure for him the confidence and good will of all who meet him;—to the hospitality of himself, and Capts. Clary and Albertis, have hundreds been indebted during the past season, for accommodations, while others took lodgings at the "Astor," owned by Mr. Childs, and managed by *Francois*, a cosmopolite, and speaking all languages, who was cook, waiter, porter, chambermaid and clerk. The "Astor" may be supposed to be a burlesque, but it is not so; and, although not a huge pile of granite like the Astor of New-York, yet it stands *on* one, and its name is quite as legitimate; and though a pine-log structure, one and a half stories high, twenty-four feet long by sixteen wide—the lower part of which is a store-house—having an addition that serves as a dining-room and kitchen, with a long table made by two boards laid upon horses;—each guest having a given space upon the main chamber floor to spread his mat and buffalo skin. This "room" is his, and his "location" as safe as if it had four walls, and a door to enter by a key "left at the office when absent;" which accommodations farther excel *all competition*, than do those of its granite namesake fronting on the Park. Its name is not a burlesque; it stands upon a point of rock not fifty feet from the wreck of the John Jacob Astor, and by way of designation was first called the "House by the Astor, or by Astor Point;" and that name, which was there fixed by a misfortune, will remain while that landmark for entering the harbor endures; and it will be known as Astor Point long after that mighty house "made by hands" in pride to perpetuate a name, shall, with its founder, have passed into other keeping, and under other governors—when death shall have removed him, and new owners in *their* pride remove his name, that "the places which knew it, shall know it no more for ever;"—then, and long after then, will the wave-tossed mariner steer his bark for "Astor Point," without knowing or caring of him whose name it perpetuates, so as it is a "landmark" that guides him to a refuge from the storm;—there, too, will the merchant stand to watch the safe entrance of his cargo, and congratulate himself upon the amount of profits, as once did his great predecessor in the trade of other climes,—whose name he knows only from the locality himself is fixed on, and whose history and exploits in trade will, having passed from tradition, be found only in the specimens of ancient history preserved by the antiquarian. Thus is it with all,—even Bonaparte's will, as he said it would, "be lost in the vortex of revolutions;" yet, its longest endurance will be with the mariner, in his anxious "look out" for St. Helena.

The Garrison at Copper Harbour is about one mile east from where the town and business must be, at the west side of the harbour, which was a very judicious selection in Capt. Clary, who superintended the building of the Garrison, to have them apart.

The Pittsburgh Company's Location, No. 4, covers Copper Harbour, extending some sixty rods west of the harbour, whose operations I shall describe among the Companies at work, as I am now endeavouring to give the reader a view of the coast.

About one mile S. W. from the west end of Lake Fanny Hoe, is a lake one mile in circumference, known as Manganese Lake, which empties into the first lake with a descent of seventy-five feet. This lake, at its outlet, has made a chasm which has opened a vein of White Spar four feet thick, below which is a vein of Manganese eighteen inches thick, and below that is another vein of the same Spar one foot thick; high hills fall abruptly to the margin of this lake, which is surrounded by a handsome sandy and pebbly beach;—the waters are transparent and very deep, and a sufficient quantity of it is discharged for manufacturing purposes.

Porter's Island is formed by a small, shallow opening, in the west end of the peninsula, which forms the western part of Copper Harbour. On this island the Mineral Agency was maintained for most of the season; bnt so great was the inconvenience, that it was removed to the Garrison, where the office is kept open during this winter by Col. McNair.

At the Garrison, also, the Commissioners, Messrs. Todd, of Ohio, and Mr. Bartlett, of New-York, held their Court of Enquiry into alleged malfeasance of certain officers of the Agency, in being interested in locations, which is strictly prohibited; and, collaterally, questions between claimants of disputed locations. It would be indelicate to give an opinion as to their report upon any of the points or matters,—which report will require, too, testimony intended or expected to be taken this winter in Washington, and probably will not transpire for a year or eighteen months; but I may say that Mr. Bartlett's candour, frankness, and good judgment, inspired confidence, and Mr. Todd's seaching inquiries, while his broad, good natured manner, left witnesses at their ease, went far to elicit facts, and convinced all that the commission was in good hands. Their report, dealing as it must with delicate matters of individual character and interest, proposing, as it probably will, reforms and changes in the laws, modes, and practices, in relation to the mineral lands, and their future disposition by Congress, will, I have no doubt, from a long acquaintance with Mr. Todd, fully sustain his reputation, not only as a lawyer, but as a statesman of expanded views, and prove him a worthy son of one of the ablest judges ever upon the bench of Ohio.

CHAPTER VIII.

Agate Harbour.—Passing by a rocky coast, composed of trap and conglomerate, against which the surf dashes, for several miles, is Agate Harbour. Its general features and shape are very similar to Copper Harbour. Its entrance is the same between large rocks some sixty rods apart, with deep water between them, and shoals just inside. There is this difference, that on entering Agate Harbour you haul up square to the east to enter the first harbuor, which is formed by a narrow point of land running east and west in the centre of the east end of the bay—the first is the north harbour. To enter the south harbour, on the other side of the point, you must keep on the course at entering, till within thirty rods of the main shore, clearing thereby a reef that puts off from the point or middle land, and then haul up east, as to enter the north or outer harbour. Either of these are as good harbours as can be wished, with clay bottom at five fathoms; there is less danger in entering Agate Harbour in a gale than Copper Harbour, because you can run into the north harbour sufficiently far to be safe from the sea, under the land, but which would not be the case if compelled to seek anchorage at the west end, as at Copper Harbour. This harbour is about three miles long and half a mile at its widest place. Its name is derived from the Agates found there by early visitors, and which are yet found in great quantities. Its shores are formed of trap and conglomerate rock, like the main shore. Its entrance is marked by two large rocks which rise several feet above the water, west of which extends a reef with small islands appearing, to the west end of the bay. High hills rise from the bay which are a continuation of those of Copper Harbour.

About one mile back from the bay is a beautiful little

Lake holding the same relation to Agate Harbour that the Manganese Lake does to Copper Harbour, but which empties its waters into the bay instead of another lake—the same scenery, sandy beach, and the water power, but not the spar and manganese veins at the former. This lake is called "Shoon-e-aw" or Silver Lake, but has never been thoroughly explored for mineral or visited by many persons because covered by lease No. 18 belonging to the New-York and Lake Superior Mining Company, who are working on the east end of the peninsula of the Harbour, where there are as many veins as they are prepared at present to work, of which I shall speak when treating of the "Companies at work."

Upon the reefs in the west part of the bay, the Trout and White Fish are easily taken in great abundance with the spear and gil-net, which is also the case at Copper Harbour.

Grand Marias—Continuing along a rocky coast as between Copper and Agate Harbours, though more broken, five miles, (but the longest five miles I ever travelled,) is the western Grand Marias. This is a large indentation in the land of about three miles in circumference,—the entrance is between rocks and too shoal for vessels, though an excellent boat harbour. The western point or side of the Marias is a level peninsula formed by the make of Grand Marias on the east and Eagle Harbour on the west. The bay is shoal and the bottom sandy—high wild grass and rushes grow on its margin covering many acres. The peninsula is level twenty feet high and may be tilled; the soil is a sandy loam. This is the location of the North-western Company, which is described among those "Companies working."

CHAPTER IX.

Eaglè Harbour. About three miles coasting, similar to that passed from Copper Harbour, brings the voyager to Eagle Harbour. This is a large bay, some four miles in circumference. Its entrance is on the west side of a large rock that appears above the water about forty rods from the east cape of the harbour. There is a reef of sunken rocks square off the mouth of the bay, but the harbour may be entered from the north-west or north-east, between the reef and the capes. The west shore of the bay must be followed, however, and at a distance of twelve to thirty rods from it is ten to twelve fathoms water entirely round it. Good anchorage is also found on the east side of the bay, just south the little island, which makes a good shelter from north and easterly weather.

This is the location of the Eagle Harbour Mining Company, on Lease No. 3, which will also be found among the "working companies." From this place a good waggon-road is made, south-west to Lease No. 6, the location of the Copper Falls Company, also a "working company." The soil in the vicinity of Eagle Harbour is a sandy loam, capable of cultivation. West of the harbour, for several miles, are sand hills covered with pines, and towards the lake a good deal of low lands and swamps that empty into Cat Harbour, an inlet between Eagle Harbour and Sand Bay. There are two or three small lakes back of Eagle Harbour, like those of Copper and Agate harbours, which empty into Eagle Harbour, affording about one mile back a good water-power.

I was at Eagle Harbour on the night of the death of Dr. Houghton, where I had an appointment to meet him, the very evening of the disaster, October 13th. I had walked from Agate Harbour to Eagle Harbour, and arrived in the afternoon of as pleasant and beautiful a day as could be desired. At sun-down the wind blew gently from the land, the sun went shiningly down, the sky was clear, and every appearance was in favour of a calm night. Half an hour after dark, a storm of wind, hail, rain and snow, came from the north-east with a suddenness and fury I have rarely seen upon the western lakes, fickle as their weather is. Comfortably seated by Mr. Sprague's fire, in a bark covered building, twelve feet square, I little thought of the calamity then befalling the country in the death of Dr. Houghton, a few miles west, whose loss may be estimated by those unacquainted with his character and services, when they know that General Cass said, on learning his death, that "Michigan had better owe ten millions." This was not said of a warrior or a statesman, but a quiet man of science and perseverance.

Ignorant, however, of the sad scene transpiring west, another, almost as fatal to life, was passing just east, which soon came to our knowledge. Our pipes had been exhausted and replenished; many sage systems of philosophy had been discussed; trap and conglomerate, copper, oxydes and sulphates had been descanted on; the storm continued to howl, and the snow stole its way with the wind in through the bark roof and chinkings, only to be instantaneously changed to a mist by the warmth of a well filled stove. Thus comfortable ourselves, knowing nought of any not so, we were just going to rest, when the door opened, and a man entered, drenched to the skin. I instantly recognised Mr. Ketchum, of New-York, whom I had arranged to accompany me to the Ontonogon in my canoe, but who rather doubting my capacity to navigate my bark, declined the journey, and I put off with a fair wind, having a gentleman going as far as Agate Harbour. Mr. Ketchum was soon followed by his voyageur, a large and powerful Indian from L'Anse, named Pickett, and brother to one of the men lost with Dr. Houghton.

It seems that after I left Copper Harbour, Mr. Ketchum engaged Pickett (who had brought potatos from L'Anse to sell,) to take him to the Ontonogon. They had left Copper Harbour that day, Pickett having his son, a lad of sixteen years, to assist him. They had left in company with Mr. Baily, and stopped at his place at the Grand Marias, and taken supper, then put out from the Marias with sail set, intending to proceed all night—Pickett at the oars and his boy steering. Thus they had passed the west point of the Marias, and were going along well, when the storm struck them from the north-east, as it did Houghton. Pickett now took in the sail with all despatch, and changed places with his son, giving him the oars, and taking the direction of the canoe himself. Mr. Ketchum sat in the bottom of the canoe. The storm increased every moment, and each wave came higher and more angry than its predecessor. The boy pulled, and Pickett put forth his herculean strength—and it was necessary. The snow and sleet came with the wind—the waves rose high and dashed and roared upon the rocks but a short distance to leeward.

Precisely so was it at that moment, twelve miles west with Houghton's boat. At every wave Pickett cheered his son to pull, assuring him there was no danger, and thereby inducing an exercise of all his energies; at the same time himself watching each coming wave, and crossing them with a precision and coolness which alone could have saved them, as they did. Mr. K. said that he spoke but once, and then asked Pickett if he could make Eagle Harbour. The Indian replied, "Oui, Monsieur,"—in the same breath shouted to his boy in Indian, and on the instant the bark canoe rose high upon a wave, then sunk into the trough, to rise still higher at the next; but father and son thus strove against the elements for an hour. At length a change—the seas were not so high—they were inside the reef at the mouth of Eagle Harbour. The Indian knew his course now despite the darkness, which was intense, and heightened by the snow. A few moments, and the angry waves were only heard expending themselves upon the rocky coast astern. The water was smooth—another moment, and the light of a window was a strong contrast in their situations, to be produced in a few moments. Then, a few rods nearer the rocks, and nothing could save them. Every pull at the oars and paddle were for life. Now they were safe, with shelter and comfort at hand. So sudden was the gale in its approach, and so powerful in its advent, that the Indian had no time to put on his coat, and when he had brought in and secured Mr. K's. baggage, he pitched their tent, and him and his boy lay down and slept without fire. In the morning they appeared as usual; Pickett had consumed a jug of brandy that was among the stores, a drink from which, and a good supper before leaving the Marias, may have contributed in no small degree to "keep his courage up" in the gale, and as that jug of brandy was sent by a friend unknown to Mr. K., but fell into Pickett's hands, who secreted it, it may have been a providential thing.

Copper Falls.—A good waggon road, is now made three miles into the interior, south from Eagle Harbour to Copper Falls, upon Lease No. 8, which belongs to the "Copper Falls Company," and which is described among the working companies. From this place this road is continued near two miles to N. Y. Lease Co. No. 31, and a trail is marked over trap and sand hills, through thickets and beaver meadows, passing by some half dozen beautiful little lakes, to Sand Bay, and then the beach, skirted by sand hills, is followed, in all nine miles to

---

CHAPTER X.

Eagle River.—The snow had fallen six inches the previous night, which made the walk to Eagle River tedious, on arriving where, I first learned the death of Dr. Houghton the previous night. Eagle River, like many other streams of this country, is wonderfully magnified by the title of river. This stream is about three rods wide, falling to within a short distance of the lake, affording in its mouth a safe place for small boats. A pier may be erected which will allow vessels to load and unload their cargoes, for stone is handy and plenty. The place is covered by Lease No. 2, which belongs to the "Lake Superior Mining Company," better known as the famous Boston Company, which will be found at the head of the list of "Working Companies."

From this place a waggon-road is also made three and a half miles to Lease No. 7, (and is being continued to Lease No. 10, by the Albion Company, who are erecting buildings,) which belongs to the Pittsburg Company, spoken of as the lessees of Copper Harbour, whose works at this place are noted with their others.

Having kept the road which runs near the Eagle River, ascending with it into the interior, and by which I had already got 300 feet above Lake Superior, I here left the valley of the Eagle River, passing the drifts into the side hill, and climbed some 300 feet, to where they had first commenced work by sinking a shaft, which was abandoned, and the drifts below commenced.

From this already high eminence, I proceeded west, ascending by irregular knobs and beaches of trap rock, for, I suppose, one-half to three-quarters of a mile in a south course, when I arrived suddenly on the top of a naked rock, about twenty feet square, which raised itself above every thing else within the scope of the eye; to the north and west was Lake Superior, "a dark blue desert, waste of waters;"—to the east was Keweenaw Bay, twenty miles distant, and away to the S. E. the Huron Mountains, sixty miles or more. The portage lake and the valleys were spread out like a map before me;—at my feet was the edge of a perpendicular precipice of 800 feet; around its base swept off to west the valley of Eagle River, in which were marked alternate Copse and Beaver Meadows. The spot where I stood was a bare flat rock, the highest peak of the Keweenaw Point, upon Lease, No. 10, belonging to the Albion Mining Company. I had heard this view glowingly described, but my imagination had formed no conception of its grandeur. I have stood upon the hills of Vermont and New Hampshire, and upon the peak of Laurel Mountain, the highest of the Alleganies—I have stood by Niagara in youthful wonder, and in manhood's realization: but the water, over whose expanse the eye flits like the lost bird, finding nought to rest on, was wanting there to add immensity of space to the scene, as at the Albion Rock; nor was there, as here, the awful precipice, with its tempting depth, that Nature's instincts teach her creatures they must shun. I have seen sights, but never one before so unconfined by hill or plane, or close horizon, where the spirit seemed set free to range at will. A *scene* only to be *felt*.

Passing a little west from here, I began to descend into a gulley in this mountain ridge, by which I reached by some steep and dangerous descents, the valley below. In the valley I found the thickets to be cedar, and on the slopes, maples, oak, ash, and saw one bass-wood, which was twenty inches through. In passing down the valley in front of the precipice, I could trace veins of spar, running both perpendicular and horizontal upon its face, at different altitudes, one of which I should judge 300 feet high. I saw when on the ridge, though below the peak, where a blast had been made, and the rock corresponded with that at the drifts of the Pittsburgh Company in the same ridge, a mile or so from it.

From Eagle River, for a few miles west, the coast continues of the same rocky character which has marked it from Copper

Harbour; in fact, for several miles round and below Keweenaw Point. One and a half miles west from here, is where Dr. Houghton, the State Geologist of Michigan, also engaged in running out the mineral lands of this region into sections and townships, and for the United States Government, in surveying mineral Leases, was drowned. I have before alluded to the loss the State of Michigan sustains in his death, and his loss to the world cannot be better described, than in the apt and truthful phrase and association, in which the following extract, from the first published account, places it.

[From the letter of E. H. Thompson, Esq., to Lucius Lyon, Surveyor General of Michigan.]

"Our country, nay, the world of science was looking with more than Argus eyes for his final report of the Geology of the Northern Peninsula. Michigan beheld his scientific talent and moral worth with pride and admiration; but alas, he has tracked the steps of glory to a watery grave!

"The transcendent genius of *Cuvier* expired in revealing the colossal and unknown forms belonging to the remote ages of past antiquity. *Champoloian* died beneath an African sun, decyphering the hieroglyphics of Egypt! *Davy* fell in the midst of enlarging the boundaries to human knowledge of natural science; and *Houghton*, like them, has fallen a martyr."

Gratiot River, six miles west of Eagle, is a larger stream than Eagle River, though running but a few miles interior.

From Gratiot River the coast continues about S. W. 16 miles to the portage, with sandy beaches, and for ten miles above the portage are high sandy cliffs, continuing to that point, from whence we went back to L'Anse, the reader will recollect to follow the coast of Keweenaw Point, and having done so, we now proceed upon our journey from the portage—nine miles to

Little Salmon Trout River, which is an inconsiderable stream, dignified with river. Its soil is sandy alluvium. Six miles further is

Grave Rods River. Similar in size to the last, but with a rocky instead of the usual sandy mouth of these streams. From this the shore runs south, forming Carver's Bay, in the bottom of which, at 12 miles, passing by

Elm River, several small nameless ones intervening between Little Salmon Trout and

Mining River, which, though larger, and running farther into the country, is of the same class of small streams whose mouths fill up with sand. Several of these smaller streams are passed, and a conglomerate shore bearing pine, hemlock, and spruce, for 15 miles, to

Flint Steel River, whose mouth is also closed with a sand bar, from which, the same character of shore, country, and timber, continues six miles to

---

CHAPTER XI.

Ontonagon River. This is one of the most important of the rivers, and discharges its waters into Lake Superior 55 miles east of the Montreal River, and about 90 from Copper Harbour. Its sources are in the south-east spurs of the Porcupine Mountains, in the vicinity of the sources of the Chippewa of the Mississippi. A great number of small tributaries of this river are outlets of small lakes, but its principal source is Windy Lake, a lake five miles from "*Lake Veux Desert*," or Lake of the Old Gardens, which is the summit lake between Lake Superior and Green Bay. This Windy Lake is five miles north of "Veux Desert," and also the source of the Menominee, which falls into Green Bay. The Ontonagon is 77 miles long. From the junction of the two branches to within nine miles of the mouth, it is rapid and shoal, but below there it is still, with a good depth of water. It has been said there was 30 or 40 miles canoe navigation above the branches. There is usually ten feet water on the bar; an island in the mouth, and the channel goes straight out on the west side of it. The banks are ten to fifteen feet high, a good soil, with timber of sugar maple, oak, pine, and birch, &c. Mr. James Paul, an old explorer and miner, has made successful experiments on this river in farming, producing wheat, corn, potatoes, and turnips, &c. There was formerly a fur company's agency here, and a settlement of Indians. The famous copper rock now at Washington was taken from its place of discovery 25 miles up this river, near which, many years since, an English mining company, impelled by the finding this boulder no doubt, sank a shaft 40 feet through clay, when they reached the red sand-stone rock, showing how little they understood the nature and geology of this region; for experience and observation has proved that though the metaliferous veins sometimes pass across the red sand rock, they have never been discovered to contain other ores than zinc and iron.

The word Ontonagon is said by an intelligent gentleman, Mr. Groveret of Mackinaw, several years United States interpreter, to mean "my bowl." That an Indian girl went to the lake with a wooden bowl for some purpose, and placing it upon the water, while her attention was drawn off, the bowl had floated beyond her reach, and to attract attention screamed Onto-na-gon, Onto-na-gon,—my bowl, my bowl, and hence the name.

There is a tradition among the Indians and voyageurs, that over the ridge of mountains, in the vicinity of the "Old Gardens," is an immense rock of native copper, far excelling any thing which has come to the knowledge of the whites. It is represented as large as "a house," and much time and labour have been expended to find it. Guides professing to know its location have been employed at large wages, who could only go *almost to it*, as was the case with a certain gentleman's guide, who, after leading him about for weeks,—and when the Major had really started to return, declared the rock was only half a day further, with a pertinacity which would have moved some men to follow him, but the Major returned *without* finding it. It is my opinion that this wonderful rock will prove very like Kid's money, (always excepting the sunken ship,) the nearer you get to *it* the further *it* is from *you*. The coast runs west, and ten miles brings you to

Great Iron River. This is the next longest river to the Ontonagan and Montreal, they being the principal rivers of this region between the two great Points of Keweenaw and Great La Point. It is close to, and follows the Porcupine Mountains, which are south-west of this river, and come close to the lake, 350 feet high, running from near the shore a direction a little west of south some 80 miles, when they intersect the great east and west ridges. Along a coast of conglomerate is

Little Iron River, whose course is almost parallel with Great Iron River, and enters the lake a short distance to the east of it. Ten miles further is Rock Island, and five miles is

Carp River, or *River to the Islands*, with several small streams just west of it,—and five miles is

Presque Isle River. This is one of the broken and rapid rivers of this region, which is describable very much as the next river, six miles further.

CHAPTER XII.

Black River. It has a small harbour of some sixty rods inside the bar; it runs on the west side, and rises in the Porcupine Mountains, some twelve miles from the lake. Its source is high in the hills, and its course is rapid and broken. Dr. Houghton's report, in speaking of this river, in connection with the Montreal and Little Pasque Isle which runs parallel with Black River and enters the lake a little east of it, says: "A greater variety of grand and beautiful scenery than that presented by some of these streams in their descent to the lake, taken in connection with the wild and rugged country, can scarcely be conceived. I was particularly struck with the great variety and picturesque views furnished by the Black River in its descent from the elevated country on the west side of the Porcupine Mountains to Lake Superior. This stream was estimated to fall about 500 feet in four miles, a constant succession of falls, chutes and rapids, which continue with so little interruption, that the waters the whole distance may be said to be a constant white foam. The stream is bounded on either side by banks from 100 to 400 feet high, sometimes sloping away gently, then again rising in mural precipices of rock, separated from each other by so short distances as to appear scarcely sufficient to permit the passage of the waters of the river. The highest fall is fifty feet." Much of this is applicable to the Dead River and western Presque Isle. Continuing along a sand stone coast, eighteen miles further, is

Montreal River. It is broad at its mouth, in comparison

with most of the others, being forty to sixty yards. It cuts through the sand rock, which is 100 feet high; and eighty rods from the lake it falls eighty feet, with a pitch of forty feet. Two miles from the lake the trap and conglomerate appear, and the river breaks through with successive falls of 180 feet, two of fifty feet each. It heads in "Lac veux Desert," from the vicinity of which rises the Ontonagon. Is rapid, but above the falls is navigable for canoes for a great distance. The soil upon this river, for twenty-five miles up, is represented of the best character, and like the Ontonagon, has timber of all kinds; also like that river, many of its branches proceed from the most beautiful little lakes imaginable—and there are many of the purest water without inlet or outlet, deep, and filled with fish. Upon its bottoms grow abundantly wild hops, grapes, and currants. There are fine fisheries at its mouth, and was formerly the site of Indian wigwams. From here the coast, sand rock and sandy, runs N.W. twelve miles to

Bad River, which comes some sixty miles from Pipe Stone Lake, and is represented by explorers and old voyageurs to be navigable for canoes eighty miles. This lake is in the second range of mountains. Its mouth is 200 feet wide, and water at the lowest stage four feet on the bar. The shores are sandy, and the hills recede. From here the shore continues N.W. for twenty miles, terminating in a broken sand point, called Chagoiemegon Point, between which, on the west and the main land, is the entrance, two miles wide, of a large bay by the same name, which is twenty miles long north and east, and seven miles wide east and west, in the bottom of which bay enters the Fish River. This bay is said to afford the best fisheries of this region, but good fisheries are so plenty, it is hard to distinguish, as all imagine those places where themselves have been successful, the *best*, and for this reason there are a great many places represented by different persons, from experience, as the *best*. In coasting, this bay is left to the south by crossing its mouth between Sand Point mentioned and the entrance of

Talking Fish River, its early name. The following traditionary account of the origin of this name, will render very simple the incongruous title, and afford a clue to many others of a similar nature in the Indian country. It is thus explained by a half breed, now seventy-five years old, who has been all his life familiar with the tales and traditions of his maternal ancestors. The story was in effect, that a great while ago, an Indian, who was a very ingenious man in new inventions, arranged a system of writing in the sand and earth, or wherever he could fix them upright, with sticks or fish-bones, which he endeavoured to perfect all the people in. The habitual, or naturally idle disposition of the Indian, however, preferred ease to studying this stepping-stone to improvement, and but very few of his people, some of the principal men, could be induced to perfect themselves in this new and mysterious art. Persevering by nature, his ingenuity struck upon a means of compelling what sloth rejected. Being the master-spirit, as well as chief of his tribe, his wisdom in council made his opinions desirable in all matters of interest among his people, and his decisions were final in all private as well as public matters. This being the case, his expedient was no less effectual than ingenious. He one day came to his lodge wet and shivering with cold, and in a few words, which grew fainter and fainter till they ceased, told of falling through the ice. He continued the next day unable to articulate words or leave his bear skin. In the succeeding days his principal men came, but not a word could he utter. After a little he was able to go forth, but could not speak. He lamented it in motions, and his people were in despair. He could hear and understand them, but could only answer in a few signs. An occasion soon occurring on which his views were deemed essential, and which, as far as he could, he gave, but he could not be understood sufficiently to arrive at his precise meaning, which he lamented in signs and grimaces, until one of his men, as he had anticipated must ultimately occur, bethought himself of the fish-bones, and reminded the chief that he could understand the "talking fish," or the language of the fish-bones. The chief appeared in doubt, as if an unthought of matter was presented; at length rising, slowly sought his bag of bones, and arranged them in explanation of his views. The scheme *took*—many of his people qualified themselves; but the use of the "talking fish" was gradually dropt after his death, and degenerated to a few heiroglyphics and symbols, as in use with all Indian nations. From him the name of Talking Fish River is said to have arisen.

CHAPTER XIII.

A large point runs north-east into Lake Superior, whose eastern shore runs N. by E. from the bottom of Chegoiemegon bay with a few points and irregularities, which, however, do not alter its general course. This is Great La Point, and in its shape and size, including the Apostles Islands, which are detachments of this point, very much resembles the Keweenaw point in size, shape and direction of its coasts, the westerly of which is N. E. and S. W. The group consists of twenty islands. A line from the bottom of the bay drawn N. E. fifty miles would touch the north-eastermost point of the north-eastern island, which is ten miles long N. and S., and three miles wide; this will be the first land discovered on the correct course up the lake bound to La Point Harbour on Madeline Island, which Madeline Island is southernmost and largest of the group, (at the south end of which is La Point Harbour,) extending twelve miles N. E. by E. and S. W. by W., and three miles wide. The northeasternmost island mentioned above as the first land, is by some called St. Michael's Island. Its north point is thirty-five miles N. E. by N. from La Point Harbour, and forty-five due N. of the mouth of Montreal River; and one hundred and five miles from Fon du Lac, the south-western extremity of Lake Superior. The coast of the N. E. end of the Great La Point, twenty miles N. W. and S. E., is broken by bays and points, between which and St. Michael's Island, is an area which is thirty-five miles N. W. and S. E. by thirty miles N. E. and S. W., in which is located eighteen of the twenty Apostles, including the N. E. half of Madeline; the excluded one, being N. W. from and near the northern extremity of the Great La Point.

Among these islands are numerous harbours, and good fisheries. At La Point harbour is an A. Fur Company's station, under the management of Mr. Borup; there is also a Mission station under Mr. Hall, an Indian agency, &c., and about two hundred and fifty inhabitants. This place is ninety miles from Fon du Lac and ninety miles from the Ontonagon. Mr. Warren who was many years American Fur Company's agent at this place, informs me that the climate is not as rigorous as is generally imagined, and that the soil is generally good, producing good wheat and the other usual crops in good quality, and that stock are always healthy. The N. W. coast of this point in the main runs N. E. and S. W. like Keweenaw, conforming to the direction of the Trap Range, of which this seems to be another uplift, and for fifty-five miles towards Fon du Lac is ruggedly broken into bays and points of constant recurrence with numbers of small un-named streams entering from the neighbouring hills very like in character but more numerous than those of Keweenaw, affording harbours, water-falls and scenery both grand and beautiful, but no streams of magnitude are found until reaching the Bois Brule.

Bois Brule, *or Burnt Wood River*, enters Lake Superior thirty-five miles due east from the mouth of the St. Louis River, at Fon du Lac, (bottom of the lake) which comes in its meanderings seventy-three miles from its source, near Upper Lake St. Croix, or Sturgeon Lake, in the great east and west range of mountains of trap, flanked on north by sand stone. The shore of the river is a sandy alluvium, as the rivers Montreal and Ontonagon, with the red sand rock, and the country of the Brule has the same general characteristics after leaving Great La Point, that marks the country of the Ontonagon and that river itself, after leaving Keweenaw Point. Its mouth is thirty to fifty feet wide, sandy, and five feet water on the bar. It surpasses all other streams in its brook trout, some of them, I have the assurance of Mr. Jacobs, weighing *ten pounds.* Its waters colder and clearer, if possible, than any of the other rivers.

This river is navigable for canoes seventy-two miles. The navigable branch does not connect, but rises near a smaller lake, or rather an immense boiling spring, which interlocks the St. Croix near by, and the voyageurs make a portage of two *pauses* to that lake, which is the summit lake. A pause is a distance which varies a little according to the length of the portage and the weight of the packs. At the portage the captain goes ahead and sets a stake to which the first packs are brought; he then goes ahead and sets a stake to which the second packs from the canoe are brought; he then goes and sets another, to which the packs left at the first pause are brought, and thus the portage is marked by pauses, generally about 80 rods. There is also *two pauses* in descending with canoes from this lake by the St. Croix river, and Lower Lake St. Croix to the Mississippi, some twenty miles below the falls of St. Anthony. These *two pauses* are caused by a rapid in the St. Croix near its exit from Upper Lake St. Croix, as I before remarked, a summit lake, the Brule falling into Lake Superior, and the St. Croix into the Mississippi. This lake corresponds with Lac veux Desert. From that the Montreal and Ontonagon fall into Lake Superior, and the Menomonie into Green Bay. Lec veux Desert being south of Keweenaw Point and Upper Lake St. Croix, south of La Point, in very nearly the same latitude, but one hundred and fifty miles apart. I am informed also by Mr. Jacobs, who has passed over to the Mississippi upon this route through Great Lake St. Croix, that there are constant mineral indications, plenty of game, though the soil upon the Brule is rather inferior; good timber, with innumerable beautiful lakes, while the westing made, counterbalances its northern position in regard to climate. At the mouth of the Brule was formerly a fur company station, which has long been removed. The bottoms of this river are at first one-fourth to one-half mile wide, increasing and rather low, but rising and widening into open ridges, with oak, elm and maple, on the lower lands. The great falls of the St. Croix are accessible to steamboats from the Mississippi.

The lake shore from here bears S. W. to the entrance of the Bay of Fon du Lac, which has usually eight feet water on the bar, and is a good harbour on the south shore, to which a long narrow strip of land comes nearly across from the north shore, and around whose eastern point, between it and the main, is reached the greater bay, or end of the lake. At the bottom of this bay, which is some nine miles in circumference, is the mouth of the St. Louis River, whose branches interlock in the N. W. with those of the Upper Mississippi.

A few miles up the St. Louis is a perpendicular fall of some fifteen feet, at which is an A. F. Co. Station, and a population of 300 of French, Indians, &c. Here enters, from the S. W. the river Fon du Lac, which rises in the mountains about fifty miles from Lake Superior. About ten miles from the mouth of the St. Louis is the portage Aux Coteaux, or portage of knives, which has ninety feet perpendicular fall. The knives, from which this name is derived, are perpendicular layers of slate, whose edges are very sharp.

From Fon du Lac, the coast of Lake Superior runs with little variation, N. E. 180 miles, to the boundary line between the United States and G. Britain, where the 48th deg. N. L. cuts the mouth of the Pigeon River, as established by a joint commission under the treaty of 1818. Due east from the mouth of Pigeon River is

Isle Royal, lying near the northern shore of Lake Superior, but within the boundary of the U. S. This is a narrow island of rock, having a length of about forty-five miles N. E. and S. W., and an average width of three to eight miles, and some of its hills have an altitude of three and four hundred feet. The geology of the island has a close resemblance to that of the mineral district upon the main shore S. W., being a continuation beyond all doubt, of the up heave of Great La Point; but the dip of the rocks is reversed, and like those of the south shore, pitch towards the lake. Nearly the whole of its N. W. side is of continuous rocky cliffs, which will scarcely admit of landing, but the southern shore has several fine bays and harbours. The northeasterly end is elevated rocks, and has spits ten to twelve miles long, with scarcely half a mile width, like the fingers half spread. It is almost destitute of soil, but has the finest fisheries. The country west of Lake Superior is mountainous for 20 miles, and some of them 1200 to 1500 feet high. This Isle Royal has been reserved by the government from locations. Yet several permits have been laid upon it.

The treaty with the Chippewas, which was consummated in 1842, gave us the mineral lands of the south shore of Lake Superior. These lands were promised to the State of Michigan in lieu of lands which fell to Ohio at the conclusion of the "Toledo war." Michigan sent Dr. Houghton in 1835, to make a preliminary examination of the country. On leaving the Great Copper Rock of the Ontonagon, in 1835, the Doctor forgot his tomahawk lying upon that rock, and when he returned there in 1843, in making his official survey, he found his hatchet—showing no one had been there in eight years. The line of Michigan, runs from the mouth of the Menomonie river where it enters Green bay, nearly N. W. to the mouth of the Western or Great Montreal river where it enters Lake Superior. The territory west of this line belongs to Wisconsin to a line drawn from the falls of St. Louis river, where the American Fur Company's station is, nearly S. E. by S. to Spirit Lake, thence by Rum river nearly S., to the Mississippi.

In addition to his other duties, Dr. Houghton was authorized to locate every 16th Section, wherever he thought proper, for school purposes. Whether he has located any or not, has not transpired.

The north shore of Lake Superior is much more rugged in its features than the southerly shore, and promises but little in the way of agriculture, for a range of low but rugged mountains skirt nearly the whole of this coast, and at several points rise abruptly to a height varying from 800 to 1,400 feet above the lake. These hills form the dividing ridge between the short streams running southerly into Lake Superior, and those taking a northerly direction into Hudson's bay; and after passing the first and most elevated range, the country is mainly made up of low granite hills, frequently destitute of timber, and intervening tamarack swamps, for a distance varying from fifty to one hundred miles, when commences the comparatively level district which extends to Hudson's bay. Nothing can be conceived which would give a more desolate character to the country than this succession of low and for the most part granite hills and swamps, and this country may be said to be a portion of the district which further west, is technically known to the Indians as the "Barren Grounds." On some portions of this shore are excellent indications of mineral, and some beautiful specimens have been brought from there.

One of the concessions wrung from the "home government" by the Provincial, has been the management of the Colonial lands of which the Provincial Parliament has now the regulation. This fact enables the Canadians to enter into the production of metals, and that government has granted permits to individuals for examining and locating twenty miles square to be held twenty-one years upon payment of a per centage of products. This is open to British subjects only, but several citizens of Detroit are said to be partners with John Prince in a location; and a company of explorers with a geologist set out from the Sault on an exploring expedition upon that side, early in November last.

Here, most patient reader, our coasting is finished. You may dispose of your canoe and calamities—doff your cotton canvass suit, your red shirt and heavy boots—take off those *inches* of beard which prevent familiars recognizing you—go into a soft bed and discover, that for several nights you are unable to sleep well; after a few days, wonder at the loss of an appetite that gave pork and hard bread so fine a relish; and, as you put the Croton to your lips, remember and long for the crystal purity of the Father of lakes, the trout and sis-kaw-it, as did certain run-aways, on a time, for the "flesh pots of Egypt." When you again join the social circle where mirth and jollity prevail, and joke and song go round: when diurnal head aches succeed nocturnal dissipations—then remember the tales and songs that whiled away the evenings by the camp-fire,—the sound and sweet sleep—the vigor and health that recompensed exercise and which made life a reality, and the world beautiful before you;—when you stand upon the colossal Exchange and mark the care-overloaded hundreds hurrying to and fro, in doubt and despair of "making good" by 3',—when you see the "Bull" of yesterday the "Bear" to-day, and both "lame Ducks" to-morrow,—when you see the ignorant and vicious gloating over accidental wealth, and prizing it only for its power of licentiousness or the immunity it will purchase;—when you see civilized "man's ingratitude to man make countless thousands mourn," you will remember

> "——the poor Indian, whose untutored mind,
> Sees God in clouds and hears him in the wind,"

and half conclude if "ignorance is bliss—'tis folly to be wise."

PART SECOND.

CHAPTER I.

GENERAL VIEW OF THE COUNTRY.

WESTERLY from Point Iroquois to Chocolate river, the country is more elevated and has a much smaller proportion of wet land, than east of it. A range of hills having an elevation ranging from 300 to 600 feet above Lake Superior, commences a little easterly from Point Iroquois, and stretches very nearly north, until the western escapement appears on the coast, giving rise to the elevated hills of which the Pictured rocks and Grand Island form a part. The outline of this range of hills has the most perfect regularity, being unbroken and uniformly covered with a dense growth of timber.

West from Chocolate river to Montreal river the physical character of the country is widely different from that of the district just referred to. This country is made up of a series of irregular knobby ranges of hills, that have a general easterly and westerly direction, with intervening valleys of flat and rolling land. These hills not unfrequently rise to a height of from 600 to 900 feet, very near to the coast of Lake Superior, and 15 to 20 miles south from the coast some points rise from 1200 to 1300 feet above the level of the Lake. The ragged outline which this district presents when viewed in detail from the Lake, contrasts in a striking manner with that of the country lying east from Chocolate river, for instead of the unbroken range of hills covered with a dense forest, there is a series of hills and knobs nearly or quite destitute of timber, and frequently abrupt and precipitous.

The only exception to the east and westerly tendency of these hills is in the range constituting the Porcupine mountains. These rise somewhat abruptly almost upon the eastern coast of Lake Superior, at a point 37 miles north-easterly from the mouth of Montreal river, and from this point they stretch inland south-west and westerly in the direction of the sources of the Wisconsin. The most elevated points of the Porcupine mountains near Lake Superior attain nearly 950, but some of the knobs inland are 1000 to 1500 feet, above the Lake. The valleys separating these ranges of hills are uniformly heavily timbered, and by far the largest proportion of the timber is beach and maple.

The length of the hilly or mountainous district, estimating in a direct line west from Chocolate River to the Montreal River, is very nearly 160 miles, and it does not probably extend at any point more than twenty or twenty-five miles south of this line, between the points mentioned. The greatest width of the district would be opposite Keweenaw Point, which extends sixty-seven miles north from this line, making the widest part about eighty-seven miles. Keweenaw Bay of Lake Superior stretches sixty miles from Keweenaw Point south, into this hilly and mountainous country.

South from the range of hilly country alluded to, and extending to Green Bay, the country at first becomes more level, and finally flat, though some regular and unbroken ranges of hills occur. In topography, it more resembles the country east of Chocolate River. Of the country between the mineral region and Green Bay but little will be known until the surveys are finished.

The Streams which discharge their waters into Lake Superior upon its southern shore, are invariably short, and, with very few exceptions, the quantity of water is small. This remark, in fact, may apply to the whole of the region of country surrounding the Lake; for this immense body of water is completely surrounded by hills, that, at no great distance from the Lake, fall away more or less rapidly. Thus, while many of the streams discharging their waters into Lake Michigan, Green Bay, and the Mississippi, have their sources near to the south shore of Lake Superior, so also many of these streams which discharge their waters into Hudson's Bay, have their sources near to the north coast of the Lake. The near approach of the summits of those ranges of hills surrounding the Lake, to the immediate coast, leaves the area of country draining into Lake Superior comparatively small.

The vicinity of the Tequoimenon River, which is the only stream east from Chocolate River that in reality breaks through the range of Sandstone hills,—before mentioned as extending from Point Iroquois,—enters the Lake about eighteen miles S. W. from White Fish Point—four to four and a half feet water on the sand bar; for seven or eight miles up, the water is ten to fifteen feet deep. Some of the sources of this stream are very near Lake Michigan. This stream, with the exception of a few rapids and chutes, is a sluggish stream, though at one point the whole stream falls perpendicularly forty-six feet—the beauty of which is enhanced by the high and overhanging rocks upon either side. This fall is at the passage, through the Sandstone hills, probably fifteen or twenty miles from the Lake. Those streams which occur between Chocolate River and Keweenaw Bay, are, with the exception of Huron—which is navigable for canoes for some distance—small; and with the exception of this and Dead River, it is believed they all have their sources in small lakes at the bases of neighbouring hills, which rarely recede farther than three or five miles from the coast.

CHAPTER II.

The following arrangement of the rocks of the Lake Superior mineral region, and their mineral contents, is condensed from Dr. Houghton's Report:

Primary Rocks.—The rocks constituting what may be considered the true primary group of this region, are chiefly granite, syenite, and syenitic granites, first seen upon the south coast of Lake Superior, constituting a rocky point known as Little Presque Isle, south-east from Keweenaw Point, near the Dead River, and known as Granite Point.

These rocks frequently appear upon the coast north-westerly, as far as Huron River, and forming the Huron Islands. West from Huron Islands, no rock appears upon the coast which in a strict sense is primary. At Huron Islands, these rocks rise upon the main land 300 to 700 feet above the lake, called "Huron Mountains," and this range of hills is continued to the south-west nearly or quite to the sources of the Wisconsin, and are joined with the south-west prolongation of the Porcupine Mountains.

Although the usual ternary compound of quartz, feldspar, and mica, occurs but rarely in the vicinity of the coast, or even in any portion of the range, nevertheless, the great mass of rocks therein may in a broad sense be called granite. The above compound is more common in the westerly than easterly portion of the range. Proceeding north upon this range, the rocks change character almost imperceptibly, the quartz being displaced with a binary compound of feldspar and hornblende, which assumes a granular structure constituting greenstone, with an intermediate, which may be called syenitic greenstone. The primary rocks which appear in the vicinity of Lake Superior, in the several ranges of hills extending from a point little south of Dead River to Huron River, are mainly syenite, or syenitic granite; is extremely compact, and the constituent minerals are mostly in small crystals, feldspar occasionally, in a more crystalline form. The granite rocks in a southwesterly direction are traversed by dykes of the greenstone, which forms the north-westerly ranges of hills. The courses of these dykes are invariably marked by striking changes in the character of the rock traversed, and in the layer of these dykes the evidences of change produced by the injection of heated matter, extend to several hundred feet upon either side of the dyke itself. There is an identity between the rock of the dyke and the ranges of greenstone north-west, in the similarity of their mineral character, and the quick succeeding dykes, till it is difficult to determine which predominates. The greenstone traverses the rock in all directions. Veins of other matter are very rare. In one instance a vein of porphyry three feet in width, itself traversed at angles of 53 and 107 degrees by greenstone of less width, was discovered, showing the greenstone most recent. The veins of greenstone traversing the granite vary from a mere line to 60 feet wide, and they waste more rapidly than the rock they traverse, which is peculiarly so upon the lake coast, leaving narrow bays and perpendicular walls, while the granite remains unaffected. Upon the north coast of the lake the granite and syenite, or syenitic granite, occasionally appear upon the shore, but more frequently these rocks are flanked on the south by the greenstone with occasional narrow bands of sandstone, thus precisely reversing the magnetic order of these rocks upon the south shore.

The term *Greenstone* is used in its generic sense, and applied to all compact rocks of a granulated structure belonging to the trap range.

MINERALS OF THE PRIMARY ROCKS.

Schorl.	Actynolite.	Feldspar.
Tourmaline.	Mica.	" Red.
Hornblende.	Quartz.	

CHAPTER III.

Trap Rocks.—Flanking the primary rocks already described on the north and northwest, are a series of ranges of hills, stretching generally in a direction south-westerly and north-easterly, which attain an altitude of from 300 to 900 feet above the lake. Along the lake side of Huron Mountains, they are less broken than the primary hills. The range of these rocks may be said to commence at the extremity of Keweenaw Point, running in a generally south-west direction, gradually receding from the coast until their crossing the Ontonagon River, nearly 25 miles from the lake. Westerly of the Ontonagon, this range becomes confounded with the northerly portions of the Porcupine Mountains, while west from these mountains a portion supposed of this same range inclines west, and approaches the coast, until, at the crossing of the Montreal River, it is but two miles from Lake Superior. West from the Porcupine Mountains, a second range of trap is continued at a distance of from 15 to 20 miles inland. The trap range of Keweenaw Point may be estimated to compose one-third of the entire width of the point, and the southwesterly portions are made up of compact greenstone, while those portions to the northwest are amygdaloid. This trap, through its whole course on the north and north-west sides, is bounded by hills of conglomerate and sand-stone, some as high as 400 feet.

The almost insensible gradations by which the granite rock passes into the greenstone of the trap formations, and the near analogy of the whole of the rocks of both formations to each other, renders it more convenient, at the same time it is more simple to follow the order adopted than their strict chronological one.

The rocks of the outer or northwestern range of hills, which were clearly of the series of the uplifts, bears more unequivocally the evidences of igneous origin, than either of the outer ranges alluded to. The rock upon the south flank of these hills is invariably very compact greenstone, while upon the north-westerly line it is almost as invariably an *amygdaloid*, or has at least an amygdaloidal structure. The causes of this difference of the structure of the rock, upon the opposite sides of this range of hills, when carefully examined, are very apparent; for it is very evident that the uplift of the rocks of the range of hills was wholly upon the south-east side, and while the rocks of this portion were in a solidified state; or in other words, that a point in Lake Superior may be regarded as the fixed axis of the uplifted mass, and sustained by the fact that the sedimentary rocks of the south or south-east are scarcely disturbed so far as it regards this range of hills, while those of the north or north-western side are invariably lifted to a high angle near the range of hills, and decreases gradually as we leave them. The sedimentary rocks, which upon the north side always dip *from* the range of trap hills, are in their close proximity to the trap inclined at angles ranging from 45 to 95 degrees. Dykes traverse these sedimentary rocks of 50 to 500 feet wide, the widest have been protruded through the strata of the sedimentary rocks, and have the general inclination. The result of these dykes occurring at short distances from the main body of trap, is that the sedimentary rocks frequently so far lose their original character as scarcely to be recognised.

The rocks of the complete north-western escapement of these hills were evidently in a state of intense ignition while in contact with the sedimentary rocks, which is shown by the evident change in the latter. The origin of amygdaloid Dr. H. believes referable to fusion of lower portions of the sedimentary rocks referred to, inasmuch that in passing south the amygdaloid disappears, and its place supplied with greenstone; and again, so intimately are they blended, that no line of demarkation is discoverable. Purely sedimentary rock is scarcely found imbedded in the amygdaloid, a circumstance though not conclusive, should not be overlooked in considering of this subject.

There is a knob of trap appears and forms Granite Point, Dead River, which is mostly greenstone, though so darkly calcined as to derive the name of serpentine rock. It possesses additional interest from the unequivocal evidence of uplift, as well as the manner of their exhibition. The cliffs of trap occupy the extremity of the point, while the neck and centre portions are made up of conglomerate or trap-tuf and sand rock resting upon the trap. The stratification of these rocks dip at a high angle, showing great disturbance, as they appear upon the coast near by in cliffs from 20 to 60 feet high. But the most curious feature is that the sedimentary rocks, for a distance of several hundred feet, have been completely shattered or broken into minute fragments, which having preserved their position were again cemented by the injection of calcareous matter, which filled the most minute fissures, so much so that a hand specimen frequently contains many hundred of those veins. This rock, like the primary, is traversed by veins of a date subsequent to the uplift.

On Isle Royal and the north shore, the same character of rock appears on the southward that occupies the north side of ranges on the south shore, and the dip answering to the axis alluded to.

MINERALS OF THE TRAP ROCK.

Quartz, Common.
Do. Smoky.
Do. Milky.
Do. Greasy.
Do. Radiated.
Do. Mamillary.
Do. Drusy.
Do. Amethystine.
Chalcedony.
Carnelian.
Jasper.

Steatite, Common.
Asbestus
Amianthus.
Calcareous Spar.
Copper, Native.
Do. Pyritous.
Do. Black.
Do. Red Oxyde of.
Do. Azure Carbonate of.
Do. Green Carbonate of.
Do Do. Ferruginous.

Agate, Common.	Lead, Sulphuret of.
Do. Fortification.	Do. Carbonate of.
Augite.	Iron, Pyritous.
Actynolite.	Do. Red Oxyde of.
Serpentine.	Do. Hydrate of.
Do. Pseudomorphius.	Do. Silicate of.
Chlorite, Common.	Manganese, Ferruginous Oxyde of.
Do. Earthy.	Silver, Native, (very rare.)

CHAPTER IV.

Metomorphic Rocks.—Flanking the primary rocks on the south, is a series of stratified rocks, consisting of talacose, mica and clay slates, slaty hornblende rock, and quartz rock, the latter constituting by far the largest proportion of the whole group. In traversing the country south-easterly from Granite Point—near Dead River, before alluded to—the granite passes almost insensibly into a serpentine rock, which has a regularly jointed structure, sometimes approaching stratification; continuing in the same direction, is found a series of hornblende slates, talacose, mica and clay slates, resting against the serpentine rocks, and still farther to the south-east, the rock becomes almost uniformly quartz. The rocks of this group dip irregularly to the south and south-east, while the cleavage of the slate is very uniformly to the north.

The rocks of the Metamorphic group stretch into the interior, in a south-westerly direction, forming the south-easterly part of the hilly region. Rocks referable to this group also occur upon the north coast of Lake Superior.

These rocks are confined exclusively to the range of hills lying upon the side east of granitic rocks, which run south-west. The outline of the Metamorphic hills is less broken than either the granitic or trap ranges, rising sometimes, however, in conical peaks, closely resembling the granitic rocks. The area of country occupied by them is less than that of the primary or trap—the average width not exceeding six or eight miles, and its extent southward is unknown.

It has been already stated that Chocolate River is the boundary on the south-east, between these and the Sedimentary rocks, and that they extend in a north-westerly direction from that stream six to eight miles, to the granite, against which they rest. The group is made up of alternating lines of talacose and mica slates, sometimes graduating into clay slates, with quartz and serpentine rocks, the quartz rock constituting by far the largest proportion of the whole mass. Their cleavage is north, to north ten degress west, with an inclination of about eighty degrees, the whole appearing to dip to south-west and south. The talacose slates and quartz rocks alternate frequently with each other, and occasionally with the *serpentine.* The quartz is distinctly granular in the main, though sometimes compact.

The rock denominated *serpentine*, bears a close resemblance to greenstone, and is essentially composed of granular feldspar and hornblende, with serpentine blended.

This rock only occurs in talacose slate, as we approach the granatic region, and may be simple lines of dykes lying parallel to the line of cleavage of the slate rocks. The Metamorphic are occasionally traversed by trap dykes.

MINERALS OF THE METAMORPHIC ROCK.

Quartz, Common	Iron, Scaly red Oxide of
" Milky	" Hæmatite
" Greasy	" Pyritous
" Tabular	Steatite
Serpentine, Common	Novaculite.

CHAPTER V.

Conglomerate.—The rock to which this term is restricted, does not occur well characterized at any point east of the district referred to as the commencement of the trap group, nor has it been noticed resting upon any of either primary or metamorphic rocks, but is invariably seen resting up the trap rock. Commencing upon the north side of the trap, at the extremity of Keweenaw Point, the Conglomerate flanks the trap, whenever that appears, upon its northerly side, as far west as the boundary of Michigan, mouth of Montreal River: nor does it stop there, but is seen at intervals as far west as the head of Lake Superior, and rests upon the trap of Isle Royale, facing south-east.

In the course of the range of the Conglomerate upon the south shore, it forms a nearly continuous range of hills, with somewhat steep escapements, but with a generally rounded outline. These hills sometimes rise to a height of from three to five hundred feet above the level of the Lake.

The Conglomerate attains a very great thickness, which is greatest at its westerly prolongation, and it gradually thins out as we proceed north-easterly; but the irregularity in thickness is so very considerable, that variations of several hundred feet are not uncommon within a few miles.

The Conglomerate rock of the south coast dips in mass irregularly to the north and north-west, while that of Isle Royale dips to the south-east. There is a *mixed Conglomerate and Sand stone* which rests upon the Conglomerate, appearing in greatest thickness upon the west flanks of the Porcupine Mountains, and disappears at Keweenaw Point.

This name is attached to the lower of the sedimentaries, which appears invariably connected with or resting upon the trap rock, not having been noticed to any extent in connection with either of the other lower rocks, for it wholly disappears as we approach the granatic or metamorphic groups. Of all the sedimentary rocks, this is the most variable in thickness, and not unfrequently does a few miles make a difference of several hundred feet. The Conglomerate rock may, without doubt, be considered as a trap tuff, which was gradually deposited or accumulated around the several conical or trap knobs during their gradual elevation, and which would necessarily occupy the complete spaces or valleys between the several irregular ranges of knobs or hills.

If we regard this Conglomerate rock in this light, we will at once perceive why the rock should be variable and irregular in its thickness. The pebbles, of which the mass of the rocks is composed, consist of rounded masses of greenstone and amygdoloid trap, of which the former make up by far the larger proportion, and scarcely a pebble of any other rock than trap enters into its composition. The pebbles vary in size from that of a pea, to several pounds weight; but the average size may be stated at one and a half to two inches in diameter. The pebbles are usually united by a mixed calcareous and argillaceous cement, more or less coloured by iron; and so firm is this union, that the most compact and tough of the greenstone pebbles, will frequently break through as freely as the cement, and crevices and narrow veins are frequently seen passing indiscriminately across the pebbles and the cement. This fact is the more worthy of notice, since the pebbles are almost without exception made up of the hardest and most indestructible portions of trap rock.

The conglomerate rock can scarcely be said to occur in such form as to be well defined in any portion of the country, excepting upon the northern flank of the outer trap range, before referred to. On the northern or outer side of Keweenaw point, the conglomerate commences near the extremity, and extends several miles westerly, forming a series of abrupt and precipitous cliffs upon the immediate shore, as also a range of well defined hills, a little in the interior, which have an elevation of from two hundred to three hundred feet. After appearing for a few miles upon the coast, this rock gradually stretches into the interior, following the line before described as the most northerly bound of the outer trap range of hills, and invariably occupying a place to the north of this range, and it may be observed, nearly or quite, continuously as far as Montreal river, which stream it crosses at a short distance above its mouth, thus making its length within the limits of Michigan, computing its southerly curve, something over one hundred and forty miles; but the rock does not cease at Montreal river, for it may be seen at short intervals in the interior as far westerly as the head of Lake Superior.

At the trap knob of Granite Point, the conglomerate is imperfectly developed, but on the south-westerly side of Isle Royale it is much more perfectly so, flanking the hills of trap upon the southerly side. The conglomerate is imperfectly stratified in masses of immense thickness, and it dips upon the south shore of the lake regularly to the north and north-west (in conformity with the variation of the trap hills in their direction) and usually at angles of thirty to eighty degrees, while upon Isle Royale and the north shore, the dip is reversed, being south and south-easterly, or in other words the rock upon all sides dips in the direction of the lake basin.

Upon the south shore a little east of Montreal river, this

rock was estimated to be 5,260 feet (nearly a mile) thick, and it wedges out or thins so rapidly, that near its eastern prolongation the estimate was 1000 feet. Its greatest estimated thickness upon the north coast was 2,300 feet.

The trap dykes of this rock, are usually parallel to the line of stratification and dip, and are from fifty to several hundred feet thick, sometimes continuing several miles. In addition there are veins of a more recent date traversing the conglomerate and the dykes always at high angles with the line of the conglomerate. These last veins, which are usually more perfectly developed near the junction of the conglomerate and trap, or for a few thousand feet on each side of that junction, are clearly seen *true veins* and are with few unimportant exceptions the only veins of this range which are *metaliferous.*

For minerals of the conglomerate rock—see "*Minerals of Conglomerate and Red Sand Stone.*"

CHAPTER VI.

Mixed Conglomerate and Sand Rock.—This rock is made up of an alternating series of conglomerate and red sand stones which rest conformably upon the conglomerate rock last described, dipping with that rock into the bed of Lake Superior. This mixed rock was not noticed upon the north side of the lake, or upon Isle Royale, but upon the south shore the rock was traced continuously for a distance of about one hundred and thirty miles, extending from a few miles westerly from the extremity of Keweenaw point, to Montreal river. It follows the line of the conglomerate before described, stretching from Keweenaw point in a south-westerly direction, and again curving to the north-west, forming as it were, a crescent, the result of which is, the rock only appears for a limited distance upon the lake shore at Keweenaw point.

From a point eighteen miles easterly of Montreal river it wedges out or thins rapidly; proceeding west, and towards the head of the lake it wholly disappears or becomes merged in the conglomerate below and sand rock above. Its greatest observed thickness was four thousand two hundred feet.

The conglomerate portion of the mixed rock consists of strata of conglomerate varying from a few feet to several hundred feet in thickness and composed of materials in all respects resembling the constituents of the conglomerate rock already described, and similarly situated.

The sand stone portion of the formation occurs in a strata of very nearly corresponding thickness, and the two rocks may be said to form nearly equal portions of the mass. But the material of which this sand stone is composed differs widely from that of the true sand rock lying above, for while the latter is chiefly made up of the quartz ore materials, the former is composed of materials bearing a close analogy in composition to those of the conglomerate rock itself; or in other words, the sand stone consists chiefly of green stone so much comminuted as, when cemented, to compose a coarse sand stone. It will thus be seen that the members of this formation differ only in the degrees and fineness of the material, and the character of this material will explain sufficiently why the true conglomerate and the mixed rocks are referable to the same origin, for the materials of the several members of the group have their origin from the trap rock, and as a whole, may perhaps be regarded as a trap-tuff.

The coarser conglomerate of the formation is scarcely separated by lines of stratification, and the strata appears usually in mass, embraced between the strata of sand stone, but the stratification of the latter rock is perfect, and it bears evidence of having been deposited in shoal water, in the very abundant, perfectly defined ripple marks which it exhibits through its complete range. No fossils were noticed in connection with either the mixed rock or conglomerate lying below it.

Dykes of green stone occasionally appear in the mixed rock, but less frequently than in the rock below. These dykes almost invariably occupy places between the strata of the rock, and correspond in position to the direction and dip of the rocks by which they are embraced, or in other words, the rocky matter composing the dykes appears to have been injected in a plane corresponding with that of the stratification of the embracing rock. As in the conglomerate below, these dykes have produced very great changes in the colour and structure of the mixed rocks bounding them on either side.

In addition to these, the mixed rock is occasionally (though less frequently than the rock below) traversed by veins or cross courses of a more recent origin than the dykes (which latter they usually cross at a high angle,) their course being usually at an angle of sixty degrees opposed to the line of bearing of the mixed rock. These cross veins are usually made up of a calcareous spar or a sub-granular lime-stone, and more rarely of some variety of quartz and imperfect trap rock, the latter usually of the amygdaloid variety.

For minerals of the mixed conglomerate and sand rock, see *Minerals of the Conglomerate, Mixed and Red Sand Stone.*

CHAPTER VII.

Red Sand Stone and Shales.—This rock and its accompanying red and gray shales occupies a much larger extent of country bordering upon Lake Superior, than any other single rock or group of rocks. It rests upon the primary and metamorphic rocks, immediately west from Chocolate River; upon the conglomerate and mixed rocks, from near Eagle River, of Keweenaw Point, west to the head of Lake Superior; upon the primary trap, metamorphic and conglomerate rocks of the north shore of the lake, and upon the conglomerate rock of Isle Royale. It is this rock which forms the basis of the level plateaus, or valleys, occupying the spaces between the several ranges of hills south from Lake Superior, and west from Chocolate River. In these last situations this rock is frequently seen undisturbed to surround the basis of isolated knobs of granite, though when near to, or in contact with knobs of trap, there are invariably evidences of very great disturbance. The rocks of this group are thickest at their westerly prolongation, thinning out as they proceed easterly.

With the exception of that portion of the coast from Point Iroquois to Grand Island, the predominating rock upon the immediate coast, both on the south and north shore, there is red sand stone, for even the primary trap and conglomerate are almost invariably skirted with it. It is over this rock that the waters are discharged at the Sault St. Mary. On both the north and south shores this rock invariably dips into the lake.

This is the chief rock that appears upon the immediate coast of the south shore of Lake Superior, and it may be said, almost the complete coast of the lake. In coasting westerly, from Grand Island to the head of the lake, one would imagine he had seen little else than red sand stone, and in fact, were he to confine his examinations to the shore alone, would see no other rock for nineteen-twentieths of the distance. It is the only rock seen, in place, from Grand Island to Chocolate River; and from Chocolate River to Keweenaw Point, embracing the complete width, of the primary, metamorphic, and trap ranges—the hills forming these groups are almost invariably surrounded or flanked at their bases with this sand rock, so that even along this portion, the hills are cut off from the lake by a narrow belt of it; and northerly from Keweenaw Point to the head of the lake, no other rocks appear upon the coast, except a few trap dykes in the vicinity of the Porcupine Mountains, and of Iron and Black rivers, and a more recent deposit of clay and sand west of Keweenaw Point. It is also the southerly side of Isle Royale.

The materials of which this rock is composed differ widely from that of the sedimentary rocks described; for while they are made up of materials clearly of trappean origin, in which is very rarely quartz, this under consideration is composed of materials, the predominating portions of which are clearly derived from the granitic and metamorphic rocks, in which quartz is abundant, though with this there is usually associated more or less sand, that has all the characteristics of the comminuted trap, constituting that portion of the mixed rock before referred to. Magnetic iron sand sometimes becomes a constituent of the red sand rock, and occasional continuous strata of several inches thickness, are almost wholly composed of this material. The components of this rock are usually cemented by calcareous matter, highly coloured by the peroxyde of iron, frequently associated with argillaceous matter.

While the chief mass of the rock is a coarse grained and somewhat compact sand rock, there are portions of the formation where there are well-formed red and gray flags, and red and green shales, forming as it were beds of a very considerable thickness, and occupying large districts of country. These red and green shales are more largely developed in that district extending from Granite Point westerly to Keweenaw Bay, and upon the south shore of Keweenaw Point, extending from the head of the bay to near the extremity of the point, and largely developed. These shales more usually occur in alternating bands of deep red and green colours, the red greatly predominating, and thus are made up of argillaceous matter of sand, the whole material being of extreme fineness.

On the south side of Keweenaw Bay, near its head, an argillaceous rock appears and extends for a short distance along the coast, which is an anomaly. The rock is evidently embraced in, or rather may be said to constitute a member of the sand stone series, but it differs widely from any other rock seen in connection with it. It sometimes appears in the form of a slate, though usually a compact strata, frequently of several inches in thickness, closely resembling indurated clay. Innumerable strata or thin layers compose the mass, being of different colours, red, gray, dark brown, alternating in the same hard specimen.

Its material possesses an extreme degree of fineness, and is so soft as readily to be cut with the knife, rendering it a material from which the Indians have long manufactured pipes. It is too soft for use in sharpening tools.

The rocks belonging to the red sand stone formation, bear the evidence of having been deposited almost universally in shoal water, for the ripple marks occur abundantly at all points where the rock takes on the decided character of sand rock, and these ripple marks may frequently be seen for many miles together, as clearly and distinctly defined as they are in many of the shoal bays. Fossils are rare in these red sand rocks.

This rock is less frequently traversed by dykes of trap than either of the rocks described, though dykes sometimes traverse the whole of the several rock formations up to and including the red sand stone. Upon portions of the north coast, where conglomerate and mixed rocks are more frequently wanting, and where the red sand stone is brought more nearly in contact with the trap, these dykes are of more frequent occurrence. It is deserving of remark, where the lower rocks are either in part or wholly wanting, the red sand stone usually becomes of a deep brown colour, and the material of which the sand is composed, gradually changes from that before described to green stone.

The Sand Rock.—It has been estimated that at its westerly prolongation, the sand rock attains a thickness of 6,500 feet, gradually diminishing to the St. Mary, where it comparatively runs out. The red also thins out proceeding southerly or inland from the coast, at a more rapid rate than the fifteen feet to a mile allowed the sand rock, as was most satisfactorily shown when connected with the several primary, metamorphic and trap ranges of hills, for all, or nearly all the valleys after passing the outer northerly range of trap hills are based upon this sand rock, and since we have every reason to believe that this sand rock was deposited in part during the gradual elevation of the several chains of hills, it would follow that over these districts which were least elevated the rock would attain its greatest thickness.

The red sand rock south of Lake Superior, as well as upon the immediate coast, dips regularly northward, while that upon the north coast dips invariably southerly, or, as has been already said of the lower rocks, this rock dips regularly upon all sides into the basin of the lake, being increased however in quantity as it approaches the primary and metamorphic ranges. The line of cleavage of some of the members of the lower sand rock and shales is frequently irregular, and opposed to the true stratification of the rock.

MINERALS OF THE CONGLOMERATE, MIXED AND RED SAND ROCK.

Calcareous Spar.	Copper, Native.*	
Quartz, Common.	" Pyritous.*	
" Milky.	" Blue Carbrnate of.*	
" Drusy.	" Green " "*	
Chalsedony (occasionally.)	" Earthy*	Iron Pyritous
Cornelian (do.)	" Black.*	" Black Ox. of
Jaspar (in conglomerate.)	Zinc, Siliceous Ox. of.	Red Ox. of
Aagate (do.)	" Carbonate of."	" Hydrate of,
		" Siliceate of.

* Chiefly in veins traversing the conglomerate.

The Upper, or Gray Sand Rock. This is the only remaining rock which separates the red sand rock from the limestone lying south. It is a gray or brownish sand rock, almost wholly composed of quartz grains usually feebly cemented with calcareous matter, in which it differs from the red sand rock, as well as in epoch of disposition, and should not be confounded with it. Besides, while the red sand rock dips regularly northerly, this gray sand rock dips as regularly southerly, conforming to the limestone resting upon it, while itself resting upon the uplifted southern edge of the red sand rock below.

From Point Iroquois it stretches westerly, in an elevated and regular chain of hills, to Taquaimenon Bay, westerly from which the shape of the coast is such that these hills do not again appear upon it, until we reach that precipitous portion of the lake coast known as the Pictured Rocks, where the effects of waves and frosts upon its feebly cemented materials have left portions in overhanging precipices, by the destruction and removal of weaker and less resisting portions, creating caverns and domes, both grand and fantastic. From the Pictured Rocks the hills composed of this stone stretch off to the south-west, passing entirely south of the primary trap and metamorphic regions to a distance unknown.

This, like the lower sand rock, abounds in ripple marks, and its line of cleavage regular and frequently opposite the line of stratification, passing over considerable districts of country. Two indistinct species of fucaides were the only fossils found. The estimated thickness at the Pictured Rocks was 700 feet, thinning out, like the others, towards the east.

Tertiary Clays and Sands.—Stratified clays and sand are seen at many points, and continue for long distances upon the coast of Lake Superior, and they are largely developed at many points in the interior of the country, which sometimes attain a thickness of 200 or 300 feet, and are spread over the less elevated portions of the district, being, in many instances, the covering of the rock forming the valleys and plateaus, and sometimes forming the lake shore.

CHAPTER VIII.

MINERAL VEINS.

[It will be necessary to keep the relation which the rocks and veins traversing them, have to each other, constantly in view.]

Veins are intersected with other veins, and sometimes with veins of other metals at both acute and right angles. Two lines approaching each other, generally have a large deposit at their confluence. It is a good sign if the branches or lodes enlarge in width or depth, but bad if they are horizontal or rising. It is a sign of a poor vein if it separates or divides into strings and sharp extremities. It is even a worse sign when a vein descends perpendicular, than when it runs horizontal. Copper will pay for working when only six inches wide, and Tin when only three inches wide, in the Cornwall mines. The richest depth for Copper Ore, in mines which have been worked, has been found to be from 40 to 80 fathoms (from 20 to 60 for Tin,) although great quantities may be raised at 80 to 100 fathoms, yet the quality decreases and the Ore is too apt to be decayed. The veins of the Cornish mines run East and West varying some 15 degrees. The veins of Lake Superior run N. E. and S. W. with slight variations.

True Veins.—The northwesterly range of hills, commencing at the extremity of Keweenaw point and stretching in a S. W. direction into the interior, are more clearly of the trappose origin than either of the other ranges, and the rock of the southerly portion of this range is greenstone, while that of the northerly flank is almost invariably either an amygdaloid or a rock approaching to toad stone.

So far as the hills lying south of this northerly range are concerned, they would appear to be as a whole, deficient in minerals and the rocks are not apparently intersected by veins or dykes of any more recent date than that of the uplift of the northerly trap hills, near the Lake.

Veins of a date posterior to the uplift of the trap rock last mentioned, are of frequent occurrence, and traverse a portion of the trap range, pass into the conglomerate and sometimes completely across the three sedimentary rocks, immediately above the trap, for an unbroken length of several miles, rarely varying more than 12 or 15 degrees from a right angle to the course of the sedimentary rocks, cutting across the dyke and conforming to the dip of the sedimentary rocks. These veins all belong to a single epoch and must be regarded as *True Veins*, and all carry the same mineral contents; and, from examinations it is confidently believed that most if not all the metalliferous veins of the upper peninsular or Lake Superior region belong to the epoch of those under consideration. It is true, native metals (particularly copper,) are found in places in the greenstone, but the quantity is small and almost always may be traced to a connection with metalliferous veins in the vicinity.

Native Copper in very thin plates was occasionally noticed as occupying the joints of the greenstone of Isle Royale, though in small quantities, but the veins so far as examined there, are less perfectly developed in their passage across the conglomerate, and very rarely contain any traces of Zinc.

In speaking of the greenstone, Dr. Houghton says, " I not only include the true greenstone, but also those altered forms of gneiss and gneissoid granite which are sometimes associated with it, while the outer or northerly portion of the same range is usually composed of an amygdaloid form of trap."

CHAPTER IX.

IRREGULAR—FALSE VEINS.

After perusing the following chapter, the reader will have perceived that the condensations from Dr. Houghton's report ceased with the last, which treated of *Regular Veins*.

I write this chapter in the hope of inducing more thorough examination and minute investigation into *Irregular Veins.* The great bane and loss in mining operations is the vast amounts expended and thrown away upon *irregular, or false veins*, which proceeds from lack of power or knowledge to decide between true and false ones. This the reader may call a " conglomerate" chapter, if he will. If the suggestions it contains shall induce investigation of the subject, their object will be fulfilled, whatever may be thought of their *philosophy* or *orthodoxy*—against any arraignment for either,

I here enter my caveat, that *one possibility is just as good as another*, in defining the probable results of an *indemonstrable* proposition.

Where veins intersect the lake's shores, they are almost invariably marked by the appearance of the white spar covering, which, in many instances, are several feet wide, and may be traced by the eye into the water thirty to forty feet in depth. Several of these wider ones occur between Copper and Eagle Harbours, in which, when the spar has been removed, boulders and ragged deposites of native copper have been found of various sizes. And from one on the conglomerate edge of the shore, on Lease 15, belonging to the Boston Company, I saw already taken out, two pieces of native copper, one weighing 800 lbs., the other some 60 lbs., which were cut off with chisels and sledge hammers from an embedded sheet, four inches thick at the place of detachment, leaving the imagination to fix its own estimate of the quantity or extent of that portion remaining in the vein. Scientific men have heretofore contended that native copper existed only, disseminated and as boulders—here, at least, it appears in a sheet, but to what extent, can only be determined by working.

In some of these veins, as at Agate Harbour, different kinds of ore, or, I think, different stages of advance from ore to native copper, are found—such as the mother of ores, glaus, green carbonate, and black sulphuret, all taken from the same vein. This vein and the one from which the native copper spoken of above was taken, are but one mile apart.

Native silver and native copper are often taken out attached to each other. Some of the veins, by the reports of analyzers, have a preponderance of silver over the copper from the same vein or rock. See Dr. Jackson's report of the Eagle River vein.

Irregular veins are of very frequent occurrence as well upon the surface as below it, which often present the best appearances, and afford specimens very likely to deceive the novice in exploring and mining. There may be injections into a crevice only, which extend but a short way in depth or length; and though well filled while the cone raised by the interior pressure continued, and during which, all the crevices were wide below in proportion to their depth, and would, had this great cone held its apex attitude, been well filled veins, no doubt leading to large deposites like *true veins;* but by the subsequent action they are not; for that cone's fallen and depressed apex is now the synclinal axis of Lake Superior, to which the strata around it dip. When the *gas* by which it had been swelled forth found vent and escaped, it carried forth in the explosion those boulders of primitive rock and native metal found in all directions, and then, following, forth rushed the conglomerate, and found its level around the trap hills, which, with fire above and fire beneath, were softened—then it was amygdaloid, greenstone and trap were blended, and then was the native copper disseminated —more in some places than in others, according to the heat and supply of the ore. The apex or cone raised by distending the earth, unsupported at the escapement of the cause, gradually settled back, sinking lower and lower as the interior heat and pressure abated, and the matter in cooling contracted, closing first upon the surface, and shutting, as they descend, the seams and crevices opened by the expansion. The closing of those seams, veins, or crevices, in this way, compressed and forced down their contents as far as they closed. Failing, however, to resume their places, in many instances, have left to unknown depths *true veins*, defined by the wall rocks, which, though showing they have been rent asunder, are also smoothed by the action of heat. In some instances, there is but one defined wall rock, while the other side is filled with native copper, disseminated throughout, as at the deep shaft of the Eagle River. In other cases, as in the Pittsburgh Company's drift, three miles back of Eagle River, both sides are workable, and the metal is disseminated in the general rock, showing that the rock had been so heated as to either take up the copper when coming in contact, or to smelt such portions of the ore as it might have possessed in its organization, the latter of which is rather sustained by the frequent occurrence of toadstone, showing that something has passed away as a gas. By this it will be readily seen why miners follow a vein that widens as it descends, even though no ore be immediately found, and discard a vein that contracts, or runs up, or even horizontal. Their experience has taught them the results t expect.

I have no doubt that we have samples of the inner coating of the earth which is next and in contact with the interior fire or latent heat, presented in the lavas thrown forth, but changed in quality and appearance by the great heat they are subjected to in their passage; a heat much more intense than that of the heated matter of the interior, which *certainly is not a flame*, but probably a dull *latent fire*, drawing sufficient of oxygen from the earth to keep it like coals imbedded in a grate, and which glow on the admission of the air.

This may reasonably be supposed to be the state of the interior fire until an active property is added. There is but one active property which can reach this fire that will make it the active cause sufficient to produce the effects presented in earthquakes and volcanoes, which property must be oxygen. There is a certainty that the supply of oxygen must come from accident too, or there would be a periodical succession of these effects, which is not the case, but whose occurrence irregularly, shows their origin to be dependant upon an accidental supply of something necessary to the effects witnessed. There is but *one* way this property necessary to combustion can be *accidentally* furnished, and it is this:—The best estimates make the earth or *shell* over this latent fire *ten miles thick*, which in proportion to the size of the globe, (some eight thousand miles in diameter) is quite a thin rind, and composed too, of particles constantly changing their position, as the globe changes its shape, through which are distributed veins of various sizes carrying water. The particles of matter forming the globe are continually changing their position, from the tendency which their weight gives them towards the equater, (like clay upon the potter's spindle,) the globe thereby enlarging at the equater and flattening at the poles, a process, whose results already have enabled the vision to range sixty miles within the polar cup; and, by the enlargement of the equator and preponderance of weight on the periphery, instituted a third, or oscillating motion north and south, known as the sun's declination. In this gradual and imperceptible movement of the matter composing this rind or skin, over the interior fire, it sometimes will occur that one of these veins of water in it will be opened upon this interior world of latent fire, and the active property necessary to combustion is thereby furnished. This vein does not furnish sufficient to extinguish, it only feeds.

A kind of inflammation commences, and the supply of oxygen in the water continuing, *active* heat and fire are produced. The disease increases and follows up the stream it feeds upon enclosing its way in its melting rocks and taking new supplies from other streams it meets with ; and thus proceeding, opens at length to some fountain, which, pouring down its waters too copious to be consumed, are changed to *steam.* That steam, increasing and expanding, must escape. The column of the entering water is exercising a hydrostatic power strong as the surrounding rocks themselves ; the steam increases and expands till the rocks split to their foundations;—the earth swells forth:—its weight compresses the steam for an instant:—the uplift settles back cracking and breaking in all directions. If these throes shall dam up or cut off the supply of water, and after a partial escapement, leave a circumscribed and steady supply, the interior inflammation of the region will have an issue called a volcano. The mountains are raised and may remain or partially settle back. But, if the supply continues upon the fire, the gas will generate and expand, the earth still continues to swell forth, opening wide interior, and raising the apex of the cone at every succeeding expansion of the increasing power, till, breaking through, the humid steam first shoots up a cloud filled with dust and ashes, followed by intensely heated matter, composed of contributions from the different strata it has made its way through to the surface ; there the oxygen of the atmosphere rushes to it and gives the outpouring mass a blazing glow. Into the vacuum thus formed, the distended cone settles back,—perhaps gradually, —may be quickly, drawing the dip of surrounding strata to an axis. This *may* be so, or it may *not* be so. I have never seen such *operations*—only the effects. Who will suggest a better way of doing a large job of strata piling, mountain and valley, earthquake and volcano making ?

Should it be that the different ores are distributed promiscuously through the interior of the earth, in horizontal layers or strata of various thickness and at different depths, we may begin to account for native metals, or those found in such purity as to be called so. These ores being soft and earthy in their beds, are carried up by the first escaping column; some ejected in the explosion as dry dust, other portions, smelted in their passage, thrown forth, cooled, form the boulders; other portions in their liquid state run into cavities or cooling in the seams remain, showing, that in their particular cases, the necessary heat and accompaniments for smelting the ore, had somewhere combined. In other instances where the combination was less perfect, other states of purity are presented, as the boulders of black oxyde, having 70 to 80 per cent copper; and the sulpturets and sulphates all indicating different degrees of perfection in the action which the ore has undergone.

The reader who is opposed to theories, may say this digression is a "foolish philosophy," and has nothing to do with mining. But from what I have seen of the mines and veins, I am of opinion it will be found worth thinking upon by all "*explorers*," and "*prospectors*."

Were this the proper place, I would like to carry this theorising a little further, and endeavour to show, as I believe, that this world has undergone, and will again, changes in its *nature*, *constitution*, and *power of production*, of not only animal but vegetable organizations, involving causes and effects, in whose history our system of geology is the record of the things of yesterday, and that the prophet Daniel's "Overturn! Overturn! Overturn!" three times repeated, had a meaning in natural philosophy which the Schools do not teach, or the Savans understand in this day, viz:—that the oscillations of the globe, from the enlargement of the equator, and flattening of the poles, will, in 12,000 years, as the Hindoos say, change its axis, which will undoubtedly be known, as it was to Noah, seven years ahead, by the immense ratio of the yearly increase of the arc described upon the heavens, and marked by the heavenly bodies, as the sun's delineation, when the consummation approaches. Also show, that Daniel's "time, times and a half time," and "restoration of all things," was a part of the same subject, and would require the time of two and a half changes of axis, or the great Egyptian cycle of 30,000 years, which would restore places and limbs of the earth to their positions of latitude and relation to the planets geocentrically, which they had occupied at the commencement of that cycle, when the earth would probably be a perfect globe, and the sun have no declination. Further, that, at each of these changes of the axis of the globe, the earth, or rind of ten miles thick, is liable to be "burned, and not consumed," by the rushing in of sufficient water and air to extend the combustion, till the rind or earth breaks in by continents, and the whole be changed by fire; when she will lose her electricity and magnetism, (from over heat, as a magnetic needle will,) and, consequently, her atmosphere and motion, and be like the moon, a cold, dead body —a sort of balance-weight receiver of the surplus magnetism of the overcharged members of the system, and giving it to the negative as they approach her in their orbits;—("that time no man knoweth, not even the angels in heaven;")— until after an "eternity," from the surplus moieties at times retained, her magnetism or life may be vicuperated, and in proportion will motion on her axis return; with motion, proportionate atmosphere will again come; with atmosphere, humidity and temperature:—these will act upon, change and decompose the surface; her eternity is passed—time and vegetation commence; stinted and sickly shrubs, at first, as on she rolls, turning her equator to the sun, in time will become trees; and magnetism is busily at work the while, assorting to their different spheres and places, the materials of the different rocks and metals, and preparing ingredients for living organic productions, *perhaps only to be originated* when the sun has no declination; and when there would be no variation of the magnetic needle, from either "terrestrial" causes in changes of axis changing the lines of circulation of the fluid: or from "celestial" causes in the varied positions of the planets in their orbits, unequally charged; and which harmonious recurrence may have formed the Egyptian cycle of 30,000 years. Those who think Moses wrote of "the beginning" without understanding his subject, or that Daniel spake without a meaning, can have little or no idea of the wisdom of the Egyptian magicians, (men of science,) whose store-house of science, learning the arts, was the pyramids, which, when opened 1300 years ago, contained innumerable specimens of now lost arts, as "glass that would bend without being broken," —the written history of the world, "sealed up to the end"—a written out system of astronomy—the names of the stars, and how they moved in their courses—the various medicines and poisons, and how they operated. For a minute account of which, see "Description of the Pyramids of Egypt, by John Greaves, Professor of Astronomy, Oxford," published 1646—which work is supposed to have originated with King James I., of England, called the "learned fool,"—who, while Prince, travelled in Egypt, Arabia and Persia, having with him Mr. Greaves, in whose name this book, containing descriptions, measurements and drawings of the pyramids, was published.

But !—my enduring reader, these are matters a long way from Lake Superior, and the Copper Mines, and this is a book of facts—not theories; yet, if you shall find yourself next season, as a Tourist or Explorer, in that land of the "mountain and flood," where Summer's sunset hours are farther up the arch of night, than in the latitude you left,—if there, far from society and social haunts, by your camp-fire musing, or climbing some mountain-peak, observe the stars wending their way through space, and then remember these suggestions upon the destiny of worlds—may they induce a wider range of thought, and

> "Lead thy will submissive to great Nature's laws—
> Admire effects—and *know* they have a *cause*.

MISCELLANEOUS.

WORKING COMPANIES.

LAKE SUPERIOR MINING CO.

Eagle River.

Charles H. Gratiot is the agent and manager. This is the Pioneer Company in working, followed closely, however, by the Pittsburgh. This company have one shaft about one mile back from the river of eighty-five feet deep with drifts from it; also five others of various depths from twelve to forty feet. They have a drift started into the hill from the bed of the creek at a lower altitude than most of their shafts, which may be made an adit if necessary in future. They have about thirty buildings in all upon their location, and one hundred and forty men and women, and children in numbers. They have a stamping machine driven by water, with sieves and washers attached. Their saw-mill is down within eighty rods of the lake. Piles of ore, estimated at twelve hundred tons, broken up, lay about and look like the prepared stone for McAdamizing. The veins are native copper disseminated, and silver is represented to be more than usually abundant in this vicinity, but in what degree or quantity over what is often found in metaliferous veins remains for working and extracting to determine. Great regularity prevails here, considering there is no law, and every thing manifests that energy, industry and capital are uniting their efforts. The width of these veins cannot be determined with accuracy—the whole depth of the deepest shaft showed but little or no demarkation, as the copper seemed distributed in all directions throughout the rock. This company have sent down 4,573 lbs. of extracted native copper. The amount sent forward is no criterion to judge their works by. They have made large expenditures and yet, although they have large quantities of broken rock ready for crushing, they did not get their machine ready till very late, and although it has cost dearly in transportation, &c.,

it does not perform the work estimated or required. But improvements will continue to be made in all these matters, and ultimate profits, I doubt not, will pay present losses, emanating from a want of knowledge and experience in the business.

The rock of these works is the amygdaloid, containing disseminated globules and leaves of native copper and silver. Sometimes silver and copper are found in the same globule. The vein is represented by Dr. Jackson to be eleven feet wide and traceable for a mile. It is my opinion that there is no clearly defined *vein* eleven feet wide—I think the whole rock of the vicinity of these shafts contains disseminated metals and may be worked, a circumstance, however, not peculiar to this particular place. I think the same fact exists on many of the neighbouring locations. But none of the locations are yet explored to any extent.

The following statements of the richness and value of the Eagle River Ore, are from Dr. C. T. Jackson :—

Analysis of 1500 grains of the Rock.—Silver from the metals, 114 grs. 49 pwt.'s; copper, 27.51; silver from the washed ore, 3.75; copper, 90.35. Amount of silver, 118.24; do. of copper, 162.86. Refined or pure silver obtained by reduction of the chloride 114.5 grs.

The analysis above detailed gives the quantity of silver in a ton of the rock—152.66—valued at $20 per lb. av. $3,053 20. A ton of the rock contains 203.57 of copper, valued at 16 cents per lb.; value of one ton of the rock, $3,036 77.

The above was Dr. Jackson's first report. The following one was subsequently made, I believe at a session of Philosophers in New Haven :—

In a ton of the rude ore as delivered by the miner at the pit bank on Eagle River there is the following per centage:

Of silver - - - - - - - -	$87,25
Of copper - - - - - - - -	42,10
Total - - - - - - - - -	$129,35

And in a ton of the ore as delivered at Boston there is $568 worth of silver and over $200 worth of copper; so that it is more properly a silver mine than a copper mine; 17 lbs. 9 oz. of the clean metal was obtained from 50 lbs. of the ore, by careful assay. 50 lbs. of copper ore gave 11 lbs. 4 oz. in large pieces of copper and silver, besides the washings; and an assay of that yielded 663 grains of pure silver, or equal to 25 2-10ths of silver to a ton of ore.

PITTSBURGH COMPANY.

The Pittsburgh Company is located at Copper Harbour, under the management of Dr. William Pettit, general agent for their several works in operation. —At this place, on Lease No. 4, on Lease No. 5, on the lake west of Eagle River, and on Lease No. 12, adjoining it.

Lease No. 4 covers Copper Harbour, Porter's Island, and Manganese Lake. They have erected seven or eight good log buildings, including storehouse, blacksmith shop, &c. The vein they are working runs west of south from the harbour, where it first appears in the conglomerate at the water's edge, some eight feet wide of spar, and thence to Lake Fanny Hoe 80 rods south. Again it appears in the mountain opposite. Upon the high ground back from the bay some 50 rods, is a shaft 14 feet deep, which seemed to be on a branch rather than the vein, the ore in it being only five or eight inches wide, and inclining much to the west. About ten rods easterly of this shaft, was one 34 feet deep, which was, instead of a regular shaft, a "lining up" of the crevice between the wall rocks, which pitched to the east. This opening was 20 feet long at top, and the walls approached at the time of their pitch within 18 inches of each other, but widened below; rather a tight squeeze to pass in the ore tubs, although they were well greased with mud. The drifting below is along this crevice, from which the famous black oxyde of the Pittsburgh Co. is taken, and from which it is raised by "Jack" & Co. for $50 per ton. Jack failing in his contract whenever a "horse" comes in, and working by the month till it is removed, when he again contracts; of which he informed me confidentially, as I was looking for a "job," which he took the liberty of inferring from my dress and use of miners' phrases, as well as willingness to go down in the "tub," which soft clothes and soft nerves are not prepared for. To be suspended in "a hole in the ground," 80 feet deep,

—to hear the click of the hammers and drills below you, and have it seem as if you never would reach the bottom, and twice as long in going up, is rather squeamish at first, but "nothing when you get used to it." A third shaft, 80 feet deep, is 20 rods further south on this vein, at the bottom of which the "drift" is more west. In reaching this depth, red sandstone was passed for a considerable portion of the descent, and from thence the drifting for the vein, which was just beginning to yield again. A gang is also now at work upon probably the same vein a little west of south from this shaft, in the mountain side across the lake, where I think the work should have been begun, and all the expenditures made in driving "drifts" at different altitudes into the mountain's side, as this company's superintendent back of Eagle River, Mr. Jennings, is now doing. From the deep shaft it is 20 rods to Lake Fanny Hoe, which is very deep, and bold banks, probably a perpendicular wall, towards which, 80 feet from the surface, the south-westerly drift is progressing, and must there stop.

Extract from Stockton's Report.

"A vein of copper has recently been opened on Keweenaw Point, near Fort Wilkins, which has already yielded several tons of ore; specimens of which have been submitted to Dr. Houghton, state geologist of Michigan, who, by analysis, has found it to contain from seventy to seventy-four per cent. of copper. Specimens of the same ore have also been subjected to analysis by Dr. M'Clintock, assayer of the United States mint at Philadelphia; and the result of the examination appears from the following extract of a letter from Dr. M'Clintock to William Pettit, M.D., who has kindly furnished it for my use:

'Having found leisure, since the receipt of your letter through Dr. Jones, to make an analysis of the *copper ore* from Lake Superior, to which it refers, it affords me great pleasure to transmit the result.

'100 parts of the ore contains of—

Silex . . 7·00
Metallic Copper 70·00
Oxygen . . 17·50

Carb. Acid, &c. 5·50 = { 3·81 carb. acid, 1·69 water, } { Or, according to Klaproth, 4·13 carbonic acid, 1·37 water.

"The mass of the ore is a *peroxide* of copper, producing a rich blue colour with aqua ammonia, which the *protoxide* fails to do. The *blue carbonate* of copper constitutes but a small portion of the specimen, and seems to dip into its interior. The carbonates always contain a portion of water, and you will therefore find the *latter* estimated with the carbonic acid, &c., 5·50 being the *absent parts*; and no trace of sulphur having been discovered, they are assumed to have been the carbonic acid of the blue carbonate, and the water necessarily associated with it.

"I send you the pure metallic copper precipitated from a solution of 50 grains parts of the ore; it weighs 35 grains parts, and is therefore equal to 70 parts in 100.

"The absence of iron, sulphur, &c., adds greatly to the value of the ore, by rendering the smelting much easier, and insuring a better article when smelted." Dated February 4, 1845.

Leases No. 5 and 12, belonging to this company, are on the lake shore immediately joining and west of Eagle River, and under the sub-management of Mr. Joseph Hussey, who has commenced building, &c., on No. 12, but to what extent he has progressed I did not learn. The veins upon this location are said to be good, corresponding with those opened in the region east and back of them, on the other Pittsburgh Lease No. 5, and Albion Lease No. 10.

Lease No. 5 is under the sub-management of Mr. R. Jennings—is three miles back from Eagle River, to which a waggon road is made, which attains an altitude in this distance of 300 feet above the lake Here are fine substantial log buildings of good size, blacksmith and cooper's shop, &c. The vein is native copper, disseminated in amygdaloid trap, and runs perpendicular with N. W. and S. E. course or nearer, perhaps, N. and S. into a prrcipitous hill 300 feet high, upon the top of which the work was formerly commenced with a shaft, but which Mr. J. abandoned on taking charge, and proceeded to open drifts at three different altitudes into the hill horizontally, which, when necessary, will be connected by ventilating shafts. The "offal rock" from these drifts is thrown off a barrow run, and the ore sent down in a chute. This vein was working very fine, three feet wide where I saw it, and presented appearances of being disseminated throughout the rock. Ore sent produced 33,577 lbs.

NEW YORK AND LAKE SUPERIOR MINING COMPANY.

At Dead River, Agate Harbour and Eagle River,—Edward Larned is general agent, and Charles Larned is the sub-agent and manager at Dead River. He has seventeen persons, fifteen men and two women. These are English and Irish. The overseer is a Cornish miner, who seems to understand his business. They have erected five log buildings, including a storehouse and blacksmith's shop, and a root house.

Their buildings and shaft commenced, are on the north end of the point near the north bay, and on the west side of the river. Their shaft is opened considerably above the lake from which the hills rise back. Mr. Larned went on with his men about the fifth of October last, and had put up his buildings and worked but five days at the shaft, when I was there, October 28th. The shaft commenced here is upon a lead vein, from which they took three tons in the five days they had worked.

The harbours of this place I have described before. The river is a rapid river, (not dead) affording beautiful falls and scenery in its course, and water power at hand, though for a short distance affording a good place for vessels, if the mouth shall have the sand bar removed. The country is hilly in the rear, with good valleys, furnishing maple, oak, ash, bass-wood, pine and birch. Of the two last they burn their charcoal. This location is represented by Professor F. Sheperd to contain copper and silver, but it has not been thoroughly explored.

Adjoining this on the west, and belonging also to the N. Y. and L. S. M. Company, covering Presque Isle, is Lease No. 20. The rock is granite as at Granite Point, thrown up through the red sand stone which appears at Garlic River three miles west in cliffs, but generally forms the plateaus between the granite hills, which seem to have thrust themselves up through it. No ores except iron and zinc are found in sand stone strata; but copper is usually found in the granite above and at their junction.

Same—Agate Harbour.—This place is also described in coasting, and is covered by Lease No. 18, belonging to this company. The sub-agent at this place is Mr. Marlet, and Mr. Hitchings conducts the mining operations. There are at this place twenty-five men, four women, a number of children, to which was added a native Irishman in the first week. They have erected since their commencement on the 29th September last, five commodious log houses, including store house and blacksmith's shop, a root house or cellar above ground, twenty-four by sixteen, and sunk one shaft forty-five feet, and another twenty-two feet. The buildings and vein are on the outer neck of land, or conglomerate rock which covers the eastern end of the harbour, near half a mile from its extremity. The vein was opened at the lake bank, by sinking a shaft, which is the forty-five feet one. The spar vein of the surface was about 18 inches in width at starting. The spar continues to its present depth, affording a green sulphuret. The wall rocks have gradually divided since passing the conglomerate of the surface, and which filled the depressed line of the vein. Quantities of what is termed the mother of ores has been raised, with green and black sulpheret and spar. About thirty rods S. W. from this shaft another is sunk upon the vein at a point where an E. and W. vein intersects. The conglomerate was passed at about nine feet, and the wall rocks of the two veins were but a few inches apart, in which was again found the mother of ores, green sulpheret and spar together; and when down twenty feet the wall rocks began to separate, and a black sulpheret was found. The E. and W. vein being about ten inches, the N. and S. about five at this point, we took a double handful of the black sulpheret, which was of dark marly appearance, and like black earth, which we put in an iron melting ladle sat upon the forge fire and covered over with charcoal; it melted and poured off (what did not escape through the pores of the ladle,) yielding a piece of copper the weight of a cent. From this, others can judge its richness as well as I can. They have here two cows, pigs, poultry, and all that is necessary for pushing their works rapidly. There are several veins on this location, and two shafts have been sunk on a vein running into the hill across the bay, but which were abandoned on commencing operations on the penisula. I think this last vein, which is in the trap rock, might be traced to the neighbouring hills, and worked by drifts to greater advantage than putting down shafts on the low altitudes. This, like all of the other locations of this company, has not been thoroughly explored. Even Shooneaw Lake, one mile back, is difficult to find.

Lease No. 31, on Eagle River, belongs to this company, and is under the sub-agency and management of Mr. Burchard. This location is on the east branch of Eagle River, and the wagon road from Eagle Harbor to Copper Falls has been extended west to Mr. B.'s place of operations. He went on with supplies and fifteen men on the 12th Oct., and on the 25th had his road made and buildings well advanced. The works on this location will be in the same range of trap hills with the Copper Falls' works. This company also hold Leases No. 32 and 17, which are next south of Copper Harbour, and bordering on the south on Keweenaw Bay, and through which the Little Montreal river runs to the Baye Bris. Neither these nor their other Leases, Nos. 19, 22, 23, at the mouth of the Montreal River, are yet worked.

THE ISLE ROYALE COMPANY.

Cyrus Mendenhall is agent and manager. They have Leases Nos. 16 and 27, east from Copper Harbour, which is No. 4. Mr. Mendenhall has been at work nearly the whole summer, with from four to seven men, and now has ten. This location is composed of conglomerate ridges next the lake, with granite breaking through in small knobs and hills. He has three small buildings near a little land-locked boat harbour cut in the conglomerate shore by the action of the lake. He has sunk at this place a shaft forty-five feet deep in the trap rock, obtaining some green sulphurets, but on the whole thus far without finding any quantity of value, but a gradual increase in the indications as the shaft descends, and which undoubtedly would lead to ore, but at what depth, perseverance alone can determine. While I was at Copper Harbour, a new vein of native copper was brought to Mr. Mendenhall's knowledge by a soldier of the garrison, on the award of Col. Todd and Capt. Albertis, who gave the soldier, after examining the vein, for finding it, $50 cash, and $500 to be paid from the first product of the vein. This vein I examined in company with Mr. Mendenhall and Geo. N. Saunders, Esq. The native copper makes its appearance in a thin sheet widening inwards. Its outer edge was the eighth of an inch, but I have a piece broke out with a hammer, three-fourths of an inch. It presents itself through a crack or vein in one of the trap knobs of a hill, 100 rods south-west from their present works, which comes through the conglomerate covering of the surface of this vicinity.

This vein Mr. M. intended to commence upon as soon as he could erect his permanent buildings, which he is doing near the eastern extremity or arm of Copper Harbour, on an inlet of which he has this season raised potatos, turnips, &c., and which will produce good hay. This company have also Lease No. 17, whose north-east corner joins the south-west corner of Lease No. 4, (Copper Harbour) which is a very hilly location, and its south-east corner cuts upon Keweenaw Bay—Lease 24 which is seven miles up the Montreal River, Lease 25, five miles up the Montreal River, Leases 28 and 29, on the Montreal; for these locations men with provisions went up very lately, merely to hold and "prospect."

EAGLE HARBOUR COMPANY.

At Eagle Harbour.

This Company's location is on Lease No. 3; Mr. Sprague is Agent and Manager of the work. He has twelve men—has been lately erecting buildings, and preparing for future operations. The vein is in the Trap River, immediately on the South shore, 100 rods west of the Harbour. Here are two or three shafts, and the deepest is 33 feet, in sinking which, he thinks he has obtained sufficient ore to pay for the work. The product is native copper disseminated in trap, and similar to that obtained at Copper Falls three miles back. Mr. S. has all that is necessary to proceed vigorously with his work this winter; there is a good water fall one mile back. They are intending to lay out a village plat here next spring, and erect a public house. It is a first rate harbour, and a beautiful site for a town. This place will probably do the commercial business for the adjacent country, including Eagle River.

THE COPPER FALLS COMPANY.

Mr. Childs is Agent and Manager of this work. Their location is Lease No. 8. It is three miles, and a good wagon road, back from Eagle Harbour. Mr. C. has ereeted six log buildings. He has turned the course of a small stream, and commenced on the vein in its bed, drifting into a hill on an altitude of 100 feet from the base, and, I suppose, 300 feet above the Lake. The product of this vein, which is two feet wide, is native copper disseminated in amygdaloid trap. The spar, and other similarities, would indicate it the same vein that Mr. Sprague is working at Eagle Harbour, or the one being worked at Grand Marias, by the North Western Company. It is from this place the road is continued by the New York Company, to their works on Lease No. 31.

THE BOSTON COMPANY.

Two miles East of Agate Harbour.

Joseph Hempstead is Agent and Manager of their works, which are on Lease No. 15. This lies upon the Lake shore, next east from Agate Harbour. He has erected five first rate commodious log buildings, and has a force of twelve men. His buildings are two miles east of Agate Harbour, and 100 rods from the Lake. The land falls from his buildings south to a beautiful little Lake, half a mile long, eighty rods wide, at each end of which, up and down the valley, are fine beaver meadows of black alluvial deposit. One mile west from his buildings is a native copper vein, from which, next the Lake, was cut and blasted off a piece of pure copper, which lay on the bank when I was there, weighing probably 800 pounds, and another of 60 pounds; the first was three feet long by eighteen inches wide, and eight inches thick. It was rather an oval form, and the wide end thickest. This Company also own Leases No. 13 and 14, on the south-eastern extremity of Keweenaw Point, but are not working them. There is a most perfect natural road from this place to Agate Harbour, running upon a conglomerate, flat ridge, smooth and level.

BOHEMIAN COMPANY.

Back of Agate Harbour.

S. Mendlebaum is the Manager and Agent of this Company, which is working on Lease No. 35. The location joins its north-west corner to the south-east corner of Agate Harbour. The works of this Company are on the head waters of Little Montreal River, in the amygdaloid trap, about half a mile south of Musquito Lake—a Lake two miles long, with a handsome island in its east end. He had a company of men, on the 20th October, making a road to Agate Harbour, and putting up buildings, and making preparations for winter's work.

Ore taken from the vein at the first blast, is certified by Dr. C. T. Jackson to yield twelve and a half per cent. copper from the crude rock of the vein ; which, he represents, as similar in all respects to the veins of the Eagle River vicinity, except that the globules are fewer and larger.

NORTH WESTERN COMPANY.

At Grand Marias Harbour.

Mr. Bailey is Agent and Manager for this Company, which is located upon the peninsula, on the eastern side of Eagle Harbour, which also forms the west side of the Grand Marias Harbour. Mr. B. had just commenced his buildings on the east side of the peninsula, when I was there on the 18th of October. He had seven men, and provided with all the necessaries to proceed to mining when his buildings should be finished. The vein on which operations will be commenced has been opened with blasts, and looks well upon the surface. The mineral is the native copper disseminated in the trap, as upon the west side of Eagle Harbour. Its width appears about the same.

The fact is, it is very difficult to determine the width of these veins of native copper disseminated, for throughout this whole region, the dissemination extends into what is generally called the wall rock, while the spar and vein stones in the conglomerate define the width.

Half a mile back from Grand Marias, is Island Lake, which is one and three-quarters of a mile long east and west, and a half to three-quarters of a mile wide, with an island in it formed of conglomerate rock, and bearing a few trees

I have left off the mineral map, the dividing line between

the west line of Lease No. 18, and the west line of Lease No. 9. My reason for so doing, is that there is a claim before the Commissioners for the Grand Marias, or a strip covering the vein, which is resisted by the Eagle Harbour Company. I am satisfied, by walking from Agate Harbour to Eagle Harbour, that there is surplus lands between these lines; but the difficulty originated in measuring from Copper Harbour to Agate Harbour, and then from Eagle River to Eagle Harbour, without closing the work between these points.

THE SUPERIOR COMPANY,

Adjoining Copper Harbour.

George N. Saunders is manager and agent for this Company. Their location is Lease No. 1, and next west adjoining Copper Harbour. There are to be seen several distinct spar marked veins, in passing this location in a boat. Mr. Saunders was preparing buildings, and had men employed for winter operations. From one of them, upon whom I can rely, I am informed the veins "prospected" upon afford excellent specimens of native copper and ores, with indications of large quantities.

ALBION MINING COMPANY.

Three miles south of Eagle River.

This Company have Lease No. 10, which is cut by the west branch of Eagle River, and a stream falling into the lake on the west. The location covers and gives name to the Albion Rock heretofore described. They have contracted for the erection of buildings and opening the road this winter, from Jennings' works on Lease 5, intersecting his road to mouth of Eagle River.

NORTH AMERICAN COMPANY.

West of Eagle River.

John Bacon is agent of this Company. Their location is on Lease No. 7, which, like those of the previous described working companies, is a three miles square permit. A west branch of the Eagle River runs through it. Their place of operations is two and a half miles west of the mouth of Eagle River, and their vein is, like all the others of this vicinity, native copper disseminated. He had a force of ten men, and was preparing his buildings on the 15th of October.

CHIPPEWA MINING COMPANY.

On the Ontonagon and Eagle Rivers.

This company is located on the Ontonagon and Eagle Rivers, as I was informed by their agent at Copper Harbour, Mr. W. H. Morell, and which I find reiterated in the following extracts from a very interesting article of that gentleman, published in the New-York Courier and Enquirer. I have made an exception in this case and put down the Chippewa Company as a working company upon his and the following representations. The others I speak from personal visit and knowledge of.

The article alluded to in speaking of the Ontonagon River says:—

"In the last eight miles of its descent to the Lake, it falls at the rate of one hundred feet or more to the mile, laying bare the trap rocks which there form its bed, and exposing to the explorer the veins that traverse them. Here numerous veins containing copper have been found, most of which have been covered by permits belonging to the Chippewa Mining Co., and in proving which they have a force now at work whose operations will be continued through the winter.

"At the promontory forming Keweenaw Point, the course of the veins is from ten to twenty-five degrees south of east; higher up the Lake, at Black River, and other localities, their course is as much, and oftener more, west of south. These veins vary in width from a mere line to several yards, and their mineral contents, whether of silver or copper, are for the most part found in a native state, although in a few instances the sulphurets, and other rich *ores* of copper, have been discovered. But it must be recollected that but comparatively little of this region has been exposed to the explorer, and it is reasonable to expect that other veins, perhaps of greater extent and richer in quality than those now discovered, may yet be developed."

The last extract corroborates not only my own observation, but the representations of nearly all explorers of the Lake Superior Mineral regions, that the great preponderance of copper is found in the native state disseminated which all analyzers thus far acknowledge, is *unalloyed with other metals.*

ENGLISH COPPER PRODUCT AND COMMERCE.

Table showing the annual average produce of the Copper mines of the County of Cornwall, England, from 1771 to 1822, and her imports.

Years.	Average number of tons of ore per year.	Average number of tons of copper produced per year.	Average amount per year for which it was sold in dollars.	Average per cent of copper the product from the ore.	Average value of the copper pr. pound. cts.	m'ls
1771 to 1775 5 years	28,749	3,449	$846,283	12	10	9
1776 " 1780 5 "	27,580	3,309	826,609	12	11	
1781 " 1786 6 "	34,354	4,122	962,380	12	10	4
1796 " 1802 7 "	51,843	5,195	2,125,046	10	I8	2
1803 " 1807 5 "	70,923	6,160	3,174,725	8	23	
1808 " 1812 5 "	70,434	6,498	2,886,835	9	12	9
1813 " 1817 5 "	82,610	7,272	2,878,723	8	17	6
I818 " 1822 5 "	94,391	7,757	3,111,811	8	17	9
The av'r'ge pr. year for 51 years	57,573	5,470	$2,101,564	9 7-8	15	2-8

Great Britain imported in 1840, 28,757 Tons of Ore, at $86 70 per ton yielding in metal 5,751 Tons which sold for		$2,493,031,90
The amount produced in G. Britian estimated by Mr. D. La Beche, is about		$6,000,000,00
Making her copper trade		$8,493,031,90
Exports pr. an. 17,777 tons at 16 cts. lb	$5,973,072,00	
Leaving her to consume	$2,219,969,90	
We import into the U. S. 1,483 tons at 16 cts. per lb.		$474,560,90

By the reported assays of our ores their probable average will be found to be 25 per cent. But suppose they shall result at a minimum average of 20 per cent. The difference between that and the average of the British ores will alone enable us to undersell them, and take the markets of their present exports to the amount of $6,000,000 per annum, and we at the same time make as great profit as *they now do*, or greater,—for while they must raise their ores 500 to 1500 feet, our best veins in my opinion, will be found to be at very advantageous altitudes in the mountains. And while they will be compelled to expend immense sums on shafts and adits, we may work upon an ascending line with inestimable saving in the cost of mining.

But if even we are compelled to descend in pursuit of the ore, the present shafts, in scarcely an instance, are troubled with permanent water. I have been down all of them—in no one did I see room for apprehending trouble from that cause. I am led to this conclusion from other reasons than its present absence. If we look upon the mineral region, and the country around it, we find that the mineral has its location in an immense range of hills, among which are innumerable small lakes, and generally very deep, an evidence that their basins are very compact, or they would filter away to Lake Superior, sometimes hundreds of feet below them, and but a short distance off. And, the bottom of Lake Superior, which forms the toe of this range of hills, is at such a depth, that but for the water these hills would be immense mountains in altitude. The rock is dry, and although hard, it is not difficult to penetrate with a drill, and its compactness amply compensates in the throw of the blast for any labour required over a softer rock. It is true, that nearly all the works have been commenced by sinking shafts on the lower levels. The reason is, that men, generally experienced miners, were engaged to commence them, who had no more knowledge, however, of how mining should be done on Lake Superior, beyond the mere mechanical operations, than the most untried noviciate. Nor is it to be wondered at, for they had never seen or heard any way except to go down, down, down, year after year, driving adits for miles under ground to get rid of entering streams from a higher altitude than the ore. They followed experience in other regions, and so did geologists and mineralogists in directing them. The latter are fast finding out that their fixed systems were adapted to other regions; and the miners are finding out that it is easier to tumble their offal down a hill, than it is to raise it up a shaft with a hand-windlass. But for the time operations have been going on, the changes already made clearly indicate what Yankee ingenuity and penetration is destined to yet accomplish in copper mining. If we look to Galena, we see them there following, ferret-like, at great depths, an inch square vein of lead to advantage. If lead can be thus worked to a profit, copper certainly can.

The mining business is, however, a peculiar one, for in the best of mines, days and weeks of labour must be exhausted without any return, in the opening and preparing, and where the vein contracts. It is a business, too, that has its daily expenses, whether there is any return or not. It requires the most particular supervision of the men. The most perfect system and promptness in all its departments is of the very last importance, and not least, is a sufficient capital necessary to sustain that system. In this, as in all other undertakings, some will fail and others succeed, having the same advantages, the results depending entirely upon their respective management. Those who have capital for permanent investment, may, with ordinary discrimination, make perfectly safe and advantageous investments with very reasonable calculations on a regular and steady business, and have the chances of immense sudden profits, which will, without doubt, in many instances, be the case ; for all that is yet known of the region is, that there are abundant evidences of copper—that so far, working has almost invariably sustained appearances.

Such, however, have been the representations, in some cases, of the qualities of metal and facilities for obtaining it, that many have imagined there would soon be an over supply. Of such, let me ask, where are the immense quantities of lead and iron consumed ?—No, there is not that person now born, who will live to see copper positively decline except with the price of manual labour, in this country any more than it will be policy to drop it below competitors to usurp the market. The uses for which it is scarcely ever employed now, and would be but for the price, are almost innumerable. Now, the market is watched, and any increase of consumption is certain to bring an increase of price, and we confine it to actual necessary purposes. Soon as we produce it in considerable quantities, we shall see how the demand will increase. We shall not then send our ships to the miners to be coppered merely to save the slight difference it now affords. Yankee ingenuity, I doubt not, is destined to show itself in improvements in producing and treating these ores, as it has in every thing else it has had presented to its universal solvent powers.

This subject of copper, is no longer to be viewed as belonging peculiarly to "*Adventurers.*" Its importance and prospective results to the nation, like other great resources of our country, demands and will receive the attention of the capitalist and statesman. We are now paying $474,560 a year for copper which we may produce ourselves; add this to the trade we may have from successful competition in $6,000,000 export of those who supply us, gives a business in commerce of millions—Compare this with our export of bread-stuffs—Let the capitalist *cypher* upon it—Let the merchant think of the balance of exchange.

GLOSSARY OF MINERS' & MINERAL TERMS.

The following glossary of terms may be useful to readers, or travellers in the mineral region:

Adit—A drain. One in Redruth is driven thirty miles under ground, being a main one.

Alluvion, or Alluvium—Recent deposits of earth, sand, gravel, mud, stones, peet, shell banks, shell marl, drift sand, &c., resulting from causes now in action, of which water is a principal agent.

Amygdaloid—A trap rock which is porous and spongy, with rounded cavities scattered through its mass, which are often filled with agates and minerals.

Argillaceous—Clayey.

Augite.—A simple mineral of variable color, from black through green and gray to white. It is a constituent of many volcanic and trappean rocks, and is also found in some of the granitic rocks.

Blend—A sulphate of zinc, a shining zinc ore.

Bowlders—Erratic gravel, lost rocks; rocks which have

been transported from a great distance, and more or less rounded by attrition or action of elements.

Basalt—One of the common trap rocks. It is composed of Augite and feldspar, is hard, compact and dark green, or black, and often has a regular columnar form.

Calcareous Rocks—Synonimous with limestone.

Calcareous Spar—Crystalized carbonate of lime.

Carbonates—Compounds, having carbonic acid, oxygen and carbon.

Chalobyte—Impregnated with iron.

Conglomerate, Crag or Puddingstone—Rocks composed of rounded masses, pebbles and gravel cemented together by siliceous, calcareous, or argillaceous cement.

Crystaline—An assemblage of imperfectly defined crystals like loaf sugar and common white marble.

Cross Course—A lode intersecting at any angle, generally throwing the vein out of its course.

Cross—The best Ore.

Crushing—Grinding ores without water.

Cunieform—Wedge shaped.

Cut—To intersect by driving, sinking or rising.

Costeaning—Discovering lodes by sinking pits in their vicinity, and driving transversely to their supposed direction.

Cofering—Securing the shaft from the influx of water by raming in clay, &c.

Core—Miners usually work but six hours at a time, consequently four pairs of men are required ; the forenoon, from 6 A. M. to 12 M., from 12 to 6 P. M. & 6 to 12 & 12 to 6.

Drift—A horizontal excavation in any direction under ground, for ore ventilation, &c.

Dead Ground—The portion of lode in which the ore is dead.

Deen—The end of a level or cross cut.

Dropper—A branch where it leaves the main lode.

Driving—Digging horizontally.

Durns—A frame of timber with boards placed behind, to keep open shafts, levels, &c.

Drill—A steel-pointed iron bar from one to five feet, struck with hammer blows, making a hole for a blast.

Elvan—Properly, clay stone.

Fast—The firm rock beneath the diluvium.

Fang—A niche in the side of an adit or shaft for an air course.

Feeder—A branch where it falls into a lode.

Flookan—A soft clayey substance which is generally found to accompany the cross courses and slides, and occasionally the lodes themselves, but when applied to a vein, means a cross course or vein of clay.

Fluke—The head of the charger, an instrument used for clearing the hole previous to blasting.

Feldspar—A simple mineral and next abundant to quartz.

Foot, or Underlaying Wall—Is the wall under the lode.

Fossils—Remains of animals and plants found buried in the earth or enclosed in rocks in different stages of change.

Gad—A pointed wedge of a peculiar form, having its sides of a porabolic figure, used in the mine for wedging off splits.

Gossan—Oxyde of iron and quartz, generally occurring in lodes at shallow depths.

Gulp of Ore—A very large deposit of ore in a lode.

Gumnies—Sands or workings.

Galena—An ore of lead composed of lead and sulphur.

Garnet—Simple mineral, usually red and crystalized. Abundant in primitive rocks.

Greenstone—A trap rock composed of hornblende and feldspar.

Horse—Is where a hard formation or the wall rock intercepts a vein.

Hornblende—A mineral of dark green or black colour.

Lodes—Cracks or fissures containing ore. They vary in course, shape and size, are generally coated with a hard crystalization; the width of a lode is known by the distance between these crystalizations.

Leap—Is when a vein disappears suddenly by diminishing in quality or quantity.

Mural Escapement—A rocky cliff with a face nearly vertical, like a wall.

Marl—By this term an argillaceous carbonate of lime is usually implied.

Marly Clay—A clay containing carbonate of lime.

Metamorphic Rocks—Stratified division of primary rocks, such as gneiss, mica slate, hornblende slate, quartz rock, &c.

Metalliferous—Containing metal or metalic ores.

Mica—A simple lightish mineral, having a shining silvery surface, capable of being split into very thin scales. The brilliant scales in granite and gneiss are mica.

Mica Slate—One of the most stratified rocks belonging to the primary class, composed of mica and quartz, the mica being in layers makes it appear to predominate.

Native Metals—Metals found in a natural uncombined state.

New Red Sand Stone—A series of sandy and argillaceous, and often calcareous strata, the prevailing colour of which is brick red, but having portions greenish gray, which occur often in spots and stripes, so that the series is sometimes called variegated sand stone. In Europe lies immediately above the coal measures.

Old Red Sand Stone—A stratified rock belonging to the carboniferous group of Europe.

Oxyde—A combination of Oxygen with another body. The term is usually limited to such combinations as do not possess active acid or alkaline properties.

Primary Rocks—Those which lie below all the stratified rocks, and exhibit no marks of sedimentary origin. They contain no fossils, and are oldest or lowest strata known. Granite, hornblende, quartz, and some slates belong to this division.

Prospecting—Examining for and partially opening veins when discovered.

Pyrites—A mineral compound of sulphur and iron. It is usually of a brass yellow, brilliant, often crystalized, and frequently mistaken for gold.

Quartz—A simple mineral, composed of silex. Rock crystal is an example.

Strata.—Layers of rocks parallel to each othe.

Secondary Strata—An extensive series of the stratified rocks which composed the crust of the globe, with certain characters in common, which distinguish these from another series below them, called *primary*, and one above them called *tertiary*.

Shaft—Is what is usually called a well. Is any required diameter, depth for ventilation, or bringing up ore. After drifting has progressed, shafts are required as ventilators, to free them from the smoke of blasts.

Slate.--A rock dividing into thin layers.

Stalactite.--Concreted corbonate of lime, hanging from the roofs of caves, and like icicles in form.

Sedimentary Rocks—All those which have been formed by their materials having been thrown down from a state of suspension or solution in water.

Serpentine—A rock composed principally of hydrated silicate of magnesia, and generally an unstratified rock.

Silex—The name of one of the few earths which is the base of the flint quartz, and most sands and sand stone.

Simple Minerals—Ore composed of a single mineral substance, while rocks are generally aggregated minerals cemented together.

Syenite and Sienite—A granite rock, in which hornblende replaces the mica.

Synclinal Axis—When the strata dips downward from opposite directions like the side of a gutter.

Tertiara Strata—A series of sedimentary rocks with characteristics which distinguish them from the two other great series of strata, the secondary and primary, both of which lie beneath them.

Transition Rocks—A series of rocks which lie below the secondary and next above the primary, and are called "transition," because they seem to have been formed at a period when the earth was passing from an uninhabited to a habitable condition, and contain a number of characteristic fossils.

Trap—Trapean Rocks—Ancient volcanic rocks, composed of feldspar, hornblende and augite. Basalt, greenstone, amygdaloid and dolomite, are trap rocks.

Tuff—An Italian name for volcanic rocks of an earthy texture.

Vein Stone—A thin crystaline coating or membrane each side of a vein.

Wall Rock—The hard rock which is almost universally found on each side of a vein, between which and the vein is the crystalization called vein stone.

NATIVE METALS AND ORES.

This list will be found to cover in some character all the specimens of the Lake Superior mines, with the exception of lead, iron, and zinc.

Gold.—Native, yellow, disseminated, in membranes and flattish grains, crystallized in cubes, seldom massive, generally the colour of brass.

Silver.–Is lightish, occurs in membranes and plates, massive and disseminated, in pieces and plates, crystallized in cubes, octangular, four-sided, rectangular, prisms, &c., and appears to belong to the new primitive (trap) rocks.

Silver, black.—Is of a bluish colour, occurs massive and disseminated.

Red Silver Ore.—Dark red, and silver grey, six-sided, resplendent.

Native Copper.—Is copper-red, often tarnished, massive and disseminated, between semi-hard and soft, little frangible, malleable, heavy, found in veins and kidneys. It fuses at 27 degrees of Wedgewood's pyrometer. The largest mass ever known of native copper is the "Ontonagon Rock," weighing 3,704 lbs., from 25 miles up the Ontonagon River, Lake Superior.

Copper Glauce.—Usually dark lead-colour, passing to blackish grey, soft, frangible, heavy, in veins.

Variegated Copper.—Colour between copper and pinch-brown, occurs massive, disseminated, in membranes, crystallized in octagons, soft, heavy, frangible, found in beds and rocks.

Copper Pyrites.—Brass-yellow, shining and soft. The best Swedish give 63 per cent.

Pyritous Copper, or *Bi-Sulphuret.*—Strongly resembles sulphuret of iron, but less pale in colour, harder, and gives fire with the steel, often variegated colours. When pure, is 30 per cent. copper, 37 sulphur, 33 iron. It is the yellow ore of Sweden, Cornwall, and Cuba.

White Copper Ore.—Colour between silver and white, a bronze yellow, occurs massive and disseminated, internally glittering, rather soft, brittle and heavy, found in veins and beds in primitive rocks.

Grey Copper Ore.—Colour most commonly steel-grey,

massive and disseminated, crystallized in various forms, partially hard, brittle, and heavy, found in new primitive (trap) and transition rocks.

Red Copper Ore.—Colour dark cochineal-red, massive and disseminated, glimmering, frangible, heavy, found the same as the grey ore.

Copper, Black.—Colour between bluish and brownish-black, occurs massive and disseminated, and as a coating to other copper ores, heavy, and yielding 40 to 80 per cent., accompanied generally by copper pyrites, is found in a black powder (podge), when massive a lustre on rubbing, is of a velvet-black.

Sulphuret of Copper, (heavy,) or vitrous copper (brochant), of which a great many varieties are known, very rich, yielding when pure about 80 per cent. of copper, often crystallized, presenting a shining lead-colour, but its colour is usually dark. It crystallizes beautifully.

Tile Ore.—Red-hyacinth colour, massive, disseminated, glimmering metallic lustre, opaque, semi-hard, brittle, heavy, frangible.

Copper, Azure.—Colour smalt-blue, frangible, disseminated, particles approaching heavy.

Copper, Green.—Colour verdigris-green, different degrees of intensity, massive, disseminated, internally shining, soft, not brittle, semi-hard, heavy.

Carbonate of Copper.—Blue and green, sometimes found in beautiful crystals, and is often valuable as a gem.

Copper, Emerald.—Colour green, crystallized in six-sided prisms, shining with vitreous lustre, translucent, semi-hard, brittle, heavy.

Dioptate.—A rare and beautiful emerald malachite, consisting of oxide of copper, carbonate of lime, silica, and water.

There is also what is called *Sulphate of Copper*, *Phosphate of Copper*, *Muriate of Copper*, *Arseniate of Copper*, of infinite variations in contents, qualities, and colours.

Copper, Mica.—Usually emerald-green, massive, disseminated, crystallized in six-sided tables, translucent, transparent, soft, light, found only in veins.

Lenticular Ore.—Colour sky-blue, passing to green, crystyllized in small flat double-sided pyramids, extremely shining, translucent, soft, rather brittle, frangible.

Olivium Ore.—Colour olive-green, seldom massive, small crystals, acute rhomboids, oblique four-sided prisms, internally glittering, adamantine lustre, soft, heavy.

Manganese.—Black and shining, granular, called magnetic iron rock, is heavy, frangible, brittle, in strata, veins.

In the Cornwall veins the sulphurets of zinc occupy the upper portion of the veins, succeeded by tin-stone, and then come copper pyrites. On Lake Superior the sulphurets of zinc and copper, and oxides of copper and iron, are similarly found.

Yellow sulphuret, pyritous copper, are the main products of the Cornwall, while native copper, the oxydes and sulphurets prevail here.

GRANTEES AND No. OF LOCATION.

The following list of applications for permits, and No. of certificate is as taken from the books of the Mineral Agency at Copper Harbour, and comprises all the entries therein on the 4th of October. But no real certainty of correctness between the list and map can be depended upon, for public books and official business I never saw in such conglomerate state as those of this Mineral Agency. Col. M'Nair, the present incumbent, assisted by Mr. Eliason, I doubt not, will look out for the future and Messrs. Todd and Bartley have charge of the past. None of the entries of 1843, 113 in number, are in this list although their number is on the map.

A

Adams R. 41
Alexander W. G. 79
Atwood H. 163
Ashley A. 164
Adams J. 183
Adams J. A. 279
Anderson J. 314
Alexander John 415
Allen. Jas. S. 459
Almy John 474
Adams James 483
Adams John Q. (Monroe, Mich.) 505
Austin W. Benton 395

B

Biddle S. B. 7
Bestor C. 10
Bush C. 16
Bickly S. W. 20
Blemot J. 32
Bates M. 53
Bates G. C. 69
Berney J. E. 150
Boyd J. J. 114
Burgess J. 131
Brewster B. H. 146
Brown J. 167
Bradley J. 170
Brown S. B. 184
Boarman S. W. 484
Bonrassan C. 194
Borrdwin N. 209
Babo C. 214
Butterfield J. 220
Bartlett J. 229
Bundtell E. 230
Brewster C. W. 239
Buckley H. J. 267
Bergman H. 274
Bartlett S. 290
Bennett W. 298
Brooks E. 303
Broadhead J. R. 304
Brinkman J. 367

Name	No.
Turburt A. D.	344
Ten Eyck Z.	410
Talcott Chas. S.	418
Tiller Jas. P. (P't)	448
Tillatson Henry A.	462
Towle Milo	479
Thomas Geo C.	482
Thurber J. S.	500
Titus Henry S.	513
U.	
Updegraf N.	320
V.	
Vouthy N.	107
Vandewater E.	329
Van Dyke J. A.	447
W.	
Wells W. J.	6
Watson J. V.	12
Whitney H.	25
Ward A. H.	33
Waterbury L.	35
Waterbury J. M.	36
Ward Wm.	40
Wickwore C.	50
Winder Jno.	52
Wilkinson John	63
Wheaton T. L.	78
Warner G. C.	108
Warder W.	129
Wheelock C. B.	157
Williams T.	160
Walner F.	190
Weymeur J.	193
Webster F.	199
Whitmore G. C.	219
Wales E. B.	222
Williams G.	255
Williams A.	256
Webb J. F.	261
Williams J. M.	269
Weber W.	270
Ward J.	288
Webb J. W.	322
Williams R.	328
Wylde J.	341
Wykoff N.	342
Warring J. B.	350
Willis E.	353
Willis C.	354
Wright G.	359
Wallen M.	380
Watson J. R.	382
Waterman N.	388
Watts Samuel	398
Williams James O.	400
White Geo. R.	407
White Wm. W.	453
Williams Benj. O.	471
Wisner O. F.	472
Weightmore R. C.	481
Wing A. E.	501

VOCABULARY OF INDIAN & FRENCH.

The following Vocabulary of Words in the Indian and French, will enable the voyageur to proceed among the half-breeds and Indians, without difficulty. They are spelled as *pronounced* by Messrs. Graveret and Rousseau, of Mackinaw and Sault St. Mary's, United States Interpreters, and Mr. Warren, of La Point.

Boo-zhoo, from the French *Bon jour*, is used throughout the whole country by Indians and others. This word all understand as "how do you do?" "good morning;" &c.---is the first salutation.

Indian.	*Fr. Spelling.*	*Fr. Pron'ciati'n*	*Definitions.*
On-eu-meek	qui, quoi, oie,	ke, koo-ah, oo	How much, for.
Ah-nien-da	oie, d'oie	d oo por oo	Where, whence.
Ah-ni-pe	que, de quoi?	ke, dai koo ah	When; of what.
Man-don	ce	suh	This.
Man-don-way-ta	cet	say	That,
Mus-ca-wa	son, bruit	song, broo-e	Noise
Ma-moon	prendre	prangdr	Take.
Nin-mene	donner	do nai	Give.
Kit-che	gros	gro	Large, Largest.
Ninne	homme	omm	Man.
Quay	femme	fem	Woman.

Que-we-zans	enfant	ang-fang,	Child,
Qua-sance	fille	fee-yuh	Girl.
Nosce	pere	payre	Father.
Qua-nosce	mere	mayre	Mother.
She-me	frere	frayre	Brother.
She-she-me	soeur	seur	Sister.
Nenne	moi	mooah	Me.
Keen	vous	vooe	You.
Scuta	feu	fuh	Fire.
Scu-ta-me-gut	chaleur	shah-lur	Hot.
She-na-ma-gut	froid	froo-ah	Cold.
Gie-me-won	pluie	ploo-ee	Rain.
Ni-be	eau	o	Water.
Nee-be	riviere	re-vyayre	Small River.
Sa-got-a-gun	Lac	lack	Lake.
Kit-che-gam-e	gros lac	gro lack	Big Lake.
Per-qua-zhe-gun	pain	paing	Bread.
Wee-os	viande	vyangde	Meat.
O-pay-a-gun	fumeur	foo-mur	Pipe.
Pos-kes-e-gun	fusile	fue-seil	Gun.
Muck-ca-ta	poudre	poodr	Powder.
Ahn-win	bals de plomb	bals des plom	Lead Balls, shot.
Mish-quaw	rouge	roohje	Red.
Muck-dta-wa	noir	noo-ar	Black
Way-gaw-quet	hache	hash	Axe.
Mo-co-mo	coteau	coo-to	Knife.
Wa-bo-wi-on	blonquette	blang-kett	Blanket.
Man-i-tou	dieu	dyuh	Great Spirit.
Kitche-ke-zick	soliel	solay-uh	Sun.
Ke-zick	lune	loon	Moon.
Kee-zhe-gut	jour	zhoor	Day.
Me-kee-zhe-gut	dimanche	de-mangshe	Sabbath Day.
Ni-be-cut	nuit	noo-e	Night.
Na-que-shine	c'nuit	se-nooe	To-night.
A-mick	castor	castor	Beaver.
Ne-kick	loutra	lootar	Ottar.
Wa-goone	renard	ruh-nar	Fox.
Ah-ne-morh	chien	shaying	Dog.
Kee-goo	poisson	poo-ah-song	Fish.
Yan	peau	poo	Skin.
Muck-wa	ours	ours	Bear.
Muck-wa-yan	peau d'ours	pooh d'oures	Bear skin.
Shoon-e-aw	argent	l'arjang	Silver, copper.
Ca-noe	barquerolle	bark-rol	Canoe
Cah-ween	non	no	No.
Na-uh	oui	oo-e	Yes.
En-dosh !	haler	ha-lai	Hallo !

Counting.

Pa-zhick	un	ung.	one
Nees-bec	deux	duhe.	two
Nis-wa	trois	troo-au.	three

Nee-wen	quatrre	khah-tr.	four
Nan-nan	cinq	saingke.	five
Go-te-was-wee	sis	seze.	ix
Che-was-wee	sept	set.	seven
Me-che-was-wee	huit	oo-itt.	eight
Chong-gus-wee	neuf	nuff.	nine
Me-tos-we	dix	diss.	ten

For counting after 10, *Che* is used before the numeral, as *Che*-pa-zhick, 11, *Che* chong-gus-wee, 19 ; and in tens *She* follows-- as Cha-pa-zhick-me-tos-we-*she*—Thirty. trante trangte. And so on.

THE PLEASURE TOUR.

When I left Copper Harbor, on the 25th October, arrangements were making for the completion this winter, of a waggon road, from Copper Harbor to Eagle River. By letters received within a few days, I learn that the contemplated road is now finished between Copper and Agate Harbors, nine miles, and rapidly progressing towards Eagle River. Preparations are making to run a stage daily between Copper Harbor and Eagle River. The workmen, I also learn, are engaged in erecting Public Houses at Copper Harbor and Eagle River, and no doubt Eagle and Agate Harbors will be also supplied.

Thus, the pleasure traveller has a new route of unequalled beauty and interest opened to him, with accommodations and comforts. From Lake Superior, he may proceed by the voyageur's canoe up the Bois Brule River, through Upper St. Croix, and down the River St. Croix to its Falls, where is a village, fine Hotel, and steamboats depart for New Orleans, or up to the Falls of St. Anthony 20 miles above the junction of the St. Croix and Mississippi. From both the St. Croix Falls and Falls of St. Anthony, millions of lumber are yearly sent down. There are at each point thriving villages. This route from Lake Superior to the Mississippi, has already been traversed by many ladies with their husbands, in journeys to and from the Fur stations, and may be performed with no more fatigue or inconvenience by any lady, than the romance of a few night's camping with a good India rubber tent; sleeping on mats, skins and blankets laid upon boughs, and eating without a table. There is one party of ladies and gentlemen organized, I learn, for a trip on this route next season, from Hartford, and no doubt many others will be met there, for, from Buffalo, in three days Mackinaw is reached, another day, Sault St. Mary, in another Copper Harbor, anothr La Point, in one day more, Fon du Lac. If the traveller shall not wish to go to Fon du Lac, or if he do, on return of the steamer, he may have his canoe or canoes launched from her off the mouth of the Brula, and entering that river two days brings him to the portage of Lake St. Croix—two days more to the Falls of St. Croix.

The canoes of any required size and convenience, and the necessary out-fits of voyageurs, and supplies of any kind desired, can be obtained at all times and on terms cheap in comparison to the usual expectations, at either St. Mary's, La Point, or Fon du Lac. The canoes may be purchased and voyageurs obtained on wages, or both canoes and men may be hired for time or voyage. Also all these may at any time be obtained at the Falls of St. Croix by travellers desiring to cross from the Mississippi to Lake Superior. Even last season dozens of canoes might be seen in a day, or the light of their camp fires at night upon this route; often far ahead would be heard coming, on the night air, the voyageurs chorus, till turning a point with arrow speed and graceful circle, on they come—they meet—the chorus ceases —a mutual "bon soir!" "bon soir!"—"adieu! adieu!"— we pass,—again the song—perhaps an hour after we hear, though far apart by the river's turns, faintly, across the bends, the whoop that interlards their song. On with the current and the oar they fly toward the frigid north, and we our way toward the equator.

STEAM & SAIL VESSELS.

Before leaving the Sault St. Mary's, I examined the vessels, which are as follows:

Julia Palmer, *steamboat* belonging to Col. W. F. P. Taylor, of Buffalo, was lying at the foot of the rapids, preparatory to being hauled over the portage. She is a staunch and very well found vessel, capable of running ten miles an hour; has accommodations for 300 passengers, and is about 280 tons burden, rigged with sails. Her main deck contains a ladies' cabin, and there is below that a steerage cabin. On the upper deck is the dining cabin, on each side of which is a tier of state rooms fore and aft, opening on the deck and into the dining room. Having a larger number of state rooms than will be likely to be required, it is intended, this winter, in making many proposed convenient arrangements, to throw two of those together, with a door opening into the third, in this way forming four or more *family rooms*, to be furnished with French bedsteads and trundle beds under them for children. Forward, on the same deck, two large rooms are to be constructed from state rooms, one for a gentlemen's room, and the other a ladies' parlour. This boat will afford every comfort to the pleasure traveller, desirable.

Independence, propeller, owned by Capt. Bristol & Co., commanded by Capt. Bristol, many years a skilful navigator of the lower lakes. She is 280 tons, a good sea vessel, and of the propeller speed. She has good cabins and accommodations, and will probably leave alternately with the Palmer.

Schooner Napoleon.—Is a new and beautiful, well arranged, found, and rigged vessel, of 180 tons, as floats any waters in the world. She was built at St. Mary's, and made one trip last fall, in which she proved herself to possess useful qualities, only equalled by her beauty and symmetry. She is owned by Oliver Newbury of Detroit, who, I have lately learned, is this winter having a similar one built at the same place, to be launched next spring.

Schooner Swallow.—Capt. John Stanord, belongs to the New York and Lake Superior Mining Company. Is being overhauled this winter, for next year's business, when whoever finds himself with "the pig let out of a bag," need have no apprehensions while there is powder for the iron pilot.

Also,

Schooners Merchant and Algonquin, of about 70 tons each--the Sis-kaw-it, Fur Trader, Chippewa, and Ocean, of a smaller class, and a small British vessel, the White Fish.

PRINCIPAL COURSES AND DISTANCES ON THE LAKE.

From	To	Offing.	Course.	Miles
White Fish Point	Picture Rocks	20 miles.	W.	100
Do.	Grand Island	20 do.	W.	120
Do.	Granite Point	20 do.		165
Do.	Michipicoten Island		N.N.W.	110
Do.	Caribou Island		N.W.	70
Do.	Keweenaw Point		N.W. half W.	173
Outer side of G. Island	Granite Point		W.	40
Granite Point	Granite Island		N.	10
Granite Island	Keweenaw Point		N. b W.	73
Do.	Caribou Island		N.E.	110
Caribou Island	Manitou Island		W. b N.	93
Copper Harbour	Isle Royal Harbour		N.W.	75
Do.	Fon du Lac	100 m. W. & 160 S.W.		260
Do.	Apostles		S.W.	150

Light Houses And Beacons.

A Light-house is much wanted at White Fish Point, Keweenaw Pt. and La Point. Beacons at least, at Copper, Agate and Eagle Harbors. Posts should be erected at the other points to which vessels may be destined, on which a light might be displayed on hearing a signal gun.

The Climate.

The general impression is that the climate of the Lake Superior Country is insupportably cold in winter. The fact is the reverse.—The snows are three feet deep as an average, but the Mercury for the past three years at Copper Harbor, sank but once below zero, then only six degrees, while the average has been four degrees above zero. The Lake never freezes more than a few miles from shore.

Mechanics And Laborers.

There is no doubt Mechanics of many kinds will next season find there, the employments of a new country in the contemplated works. Wages this winter are—for best miners, $30 per month and found—second rate and laborers, $18—Blacksmiths, $40. "*Tributors*" can always obtain veins or sections to work, for a liberal per centage.

A New Crusher.

As Crushing is to be a very considerable item in producing Copper, it is proper to mention here, that a new machine for the purpose is already preparing of economical construction, by Mr Rawdon, of the firm of Rawdon, Wright & Hatch, for crushing and grinding the ore rock preparatory to washing and smelting. The inventor believes this Machine will supercede all others in its compactness, perfection of workmanship and rapidity of operation. Mr. R. asserts that it will be capable of preparing three tons per day driven by a two horse steam or other motive power. Its size—three feet by four, will render its transportation trifling. Four of them may be worked on a space twenty feet square, requiring but one man to the four. Its perfection of construction leaves but little liability to breaking or repairs. They may be so arranged that the washers will receive the pulverised ore direct.

ENGLISH COPPER STOCKS.

The result of investments in English copper mines may be judged in a degree by the following list of Copper stocks, taken at random from the price current of the London Miners' Journal of November 15, 1845.

Shares.	Company.	£Paid.	£Price.	Shares.	Company.	£Paid.	£Price.
235	Andrew and Nangiles,	25½	70	1024	Wheal Maria, - -	1	700
100	Botallack, - -	175	400	96	Tresavean, - -	10	300
114	Charlestown, - -	—	240	256	Trenow Consols, -	—	170
1000	Carn Brea, - -	15	80	5000	Treleigh Consols, -	6	3½
256	Caradon Consols, -	45	140	4000	United Hills, - -	5	5
256	Caradon Copper Mine,	4½	6	100	United Mines, -	1000	900
1000	Copper Bottom - -	1	5	6000	Wicklow Copper, -	5	18
512	Fowey Consols, - -	—	80	128	Wheal St Andrew, -	65	20
244	Grambler & St. Anbyn,	—	55	127	Wheal Virgin, - -	—	20
128	Hallenbeagle, - -	—	50	256	West Caradon, -	40	375
160	Levant, - - -	—	150	3845	West Wheal Jewel,	10½	4
128	Par Consols, - -	—	500	128	Wheal Providence,	16	120
800	South Towan, - -	10	1½	256	Wheal Sisters, - -	22½	78
128	South Caradon, -	5	400	256	West Wheal Treasury,	12	12
120	Trethellen, - - -	5	100	256	West Wh. Friendship,	—	5

A visit by Mr. G. W. Hughes to the interior of a CORNWALL MINE.

The following description of the interior of a Cornwall Mine, from the Report of Capt. G. W. Hughes, who reported to our government his examination of these mines, will be read with interest. The extracts from the same upon smelting and refining, will be interesting to all who have any wish know the process by which copper is produced from the ores, by the most approved methods:

" Before diving into the bowels of the earth, we examined the map and sections of the mine, (each gallery and pit being carefully laid down on a large manuscript map, as soon as they are projected,) the machinery of the water wheels, steam engines, &c.

" After finishing the examinations above ground, we prepared for the descent, by taking off every article of dress, and putting on clean dry flannel, coarse shoes, without stockings, and clean cotton caps on our heads, surmounted with low crowned felt hats. In this travesty, with a lighted candle in hand, and another as a reserve suspended from the neck, we were ready for the adventure.

" Now imagine a circular opening in the ground, of the diameter and general appearance of a common-sized well, with a curb to it, and the top of a ladder projecting three or four feet above the surface. On this ladder you place yourself, and commence the descent. The last advice given me as I disappeared from the open day, in reference to my personal deportment, was to be certain that one foot was safely and firmly planted on the ladder before the other was moved—a precaution on which one's life might depend. Between the thumb and fore finger of the left hand you carry the lighted candle, and with the other fingers and the right hand grasp firmly the iron rounds or steps of the ladder. These ladders are from twenty to sixty feet in length, placed at almost every degree of inclination, from the vertical to 45 degrees. In this wise is the descent made for more than 1200 perpendicular feet. At the foot of each ladder is a narrow platform on which you may rest; and it is from these platforms the lateral galleries are carried.

"Following the lode, galleries have been worked for great distances, or are in progress, on the levels along the strike of the lode; and several main pits, such as I have just described, are sunk for the convenience of the workmen. We saw the miners in all directions, on the different levels, at work, blasting and wheeling out the ore. The effect is sometimes quite striking; you see a light, twinkling like a star, apparently at a great distance; you attempt to approach it; sometimes you are half knee deep in water; sometimes you may walk erect, at others you must stoop low, and again scramble over the rubbish on your hands and knees; at times you lose sight of this polar star, towards which your course has been directed, and anon, as you pass the projection of some jagged rock, it bursts again upon your sight. Perhaps, as you pass the dark and frowning mouth of some cross gallery, you are startled by the quick sharp sound of a sudden avalanche of rocks, rushing to a lower level, or you may hear a noise like distant rumbling thunder, which you may at first take for the approach of an earthquake, and not feel very comfortable at the thought, but it is only the reverberation of the noise of a blast.

"Whilst under ground, I stopped frequently to converse with the miners, who seemed to be generally clever but not educated men. They were mostly Chartists, and ever ready for a political discussion. As they knew me to be a stranger, having no connection with the works, they probably spoke without reserve, and expressed their true opinions. One of them, whilst enumerating the working men's grievances, suddenly sprung up, and in the heat of enthusiasm struck with his pickaxe several blows into the ore, in quick succession. 'There,' said he, addressing himself to me, 'I have struck five blows, and how many think you were for myself?' I replied, I could not tell. 'Well.' he continued, 'one was for the *government*, in the formation of which I have no choice; one for the *church*, which I never enter, being a dissenter; one for the *lord of the manor*, whose face I never see; one for the *tavern keeper*, whose face I see too often; and the other for myself.'

"On returning, we came up by one of the old and disused inclined planes, partially blocked up with rubbish Sometimes we were forced to crawl on our hands and knees, and in this manner we toiled our weary way up to the adit, about sixty feet below the surface, which we reached by means of a ladder, The adit is a subterranean canal, frequently of great length, for the discharge of the water which has been pumped up from the bottom of the mine. The pumps are placed alongside the ladders, and are made very strong, to resist the enormous pressure to which they are subjected in raising the water to so great height. At the different levels, where the valves are placed, the pumps are made of cast iron, carefully screwed together, with felt between the pieces, to render them water-tight. When we reached the surface, I required the assistance of the guide to gain the platform, so completely was I exhausted by the unusual action to which the muscles of the limbs had been subjected; and for several days afterwards, I could not move without great difficulty and pain. And, speaking from experience, I would advise no one who can avoid it to make in the winter any such descent in the mere spirit of adventure.

TREATMENT OF COPPER ORES.

The treatment of different ores should vary with their qualities, but the processes are essentially the same; but, as the limits of this paper will not permit me to specify all the modifications in detail, I shall restrict myself to designating the operations to which pyritous ores, ores containing but a small proportion of sulphur, and gray or argentiferous copper, are subjected in different countries.

The treatment of the first two varieties begins with a sorting by hand, which consists in putting aside all the fragments as large as hen's eggs, and from amongst them to separate and reject the stony pieces. Those which remain are broken according to their size, and all are reduced to fragments not larger than walnuts. Another selection or picking over is then made, for the purpose of sorting them according to their richness, and rejecting those which are altogether stony. The others form three qualities, viz: 1st, fragments of pure ore; 2d, fragments slightly intermixed with foreign matter; and, 3d, fragments of the poorest kind. This sorting is more or less rigorous, according to the expense of the subsequent treatment, consisting of the price of fuel, manual labour, transportation, &c.

No. 1 is broken on a cast-iron bed, by means of a batte or flat beater, (formed of a piece of iron, six inches square and one inch thick, adapted to a wooden handle,) into fragments not larger than filberts. The selection and breaking are usually done by women and children. When this is completed, the ore is ready for calcination. No. 1 is called *prills* in Cornwall. No. 2 *the dredge ore*, (as it is termed,) is spalled by the batte, and then sent to the workshops to be riddled (sifted) and washed. No. 3 (the *halvans* or leavings) is sent to the crushing mills.

BREAKING AND SCREENING ORES.

In Cornwall, the ores at many establishments are broken by means of crushing mills, instead of being broken by battes. As the crushed ore falls from the crushing rollers or cylinders, it is received in an inclined cylindrical sieve, which, by turning on its axis, permits the larger sized ore to slide down to the bottom, while the smaller pass through its meshes. The pieces which do not pass through the sieve are returned to the crushers.

The small pieces of ore separated in the first selection, and from which all fragments larger than hen's eggs were sorted, are then sifted on an iron wire riddle; the intervals between the wires are about five lines. A workman, by repeated shaking of the riddle, filled with ore, and immersed in running water, separates it into three parts; 1st. The fine particles, which are carried off by the current, and deposited in a distant portion of the pool. 2d. The fragments, which are deposited under the riddle. 3d. The largest pieces, which remain on the riddle. The last are spread out on a table, and assorted by hand. The fragments not rejected as too poor are then broken, and resubmitted to the above named operations.

The ore deposited under the riddle is resifted, and the No. 2 (broken when the separation into three classes was made) is added. The meshes of the sieve are from twenty to thirty to the square inch. The workman who manages it shakes it by two handles, with a compound motion, both round and vertically, immersed in a shallow pool of water, above which it is never lifted. The finest particles pass through the sieve; those that remain above it are divided, by the difference of their specific gravities, into three parts. The lightest, which is above, contains so little metal that it is thrown away; the middle portion is sent to the stamping mills; and the lower part, the richest of all, accumulated during two or three siftings, needs only to be placed on a slightly inclined plane, and to be stirred with a rake while it is washed by a small stream of water flowing gently over it. This operation, which is called "jigging," is at most of the well-regulated Cornish mines performed by machinery. A rectangular sieve is suspended in a box of the same shape, filled with water, and a vibratory motion is communicated to the sieve by a working rod, connected either with a water wheel or a steam engine, as may be most convenient. The old method is still followed on the continent, and may still be seen in England at the smaller mines.

The fine ore passed through the sieve is separated into two parts by stirring in water. The richest, fallen to the bottom, requires but one washing. Finally, the different classifications of ores sent to the crushing and stamping mills, and separated from the non-metalliferous substances with which they were mixed, by stirrings and washings on inclined planes, constitute the *prepared ores*, ready for sale and reduction; while the stony parts, which always retain some ore, are thrown away.

ASSAYING COPPER ORES.

Mr. De la Beche observes "that the mode of assaying copper ores in Cornwall is usually conducted in a somewhat rough manner, and accurate results can scarcely be expected from it. Indeed, chemistry has as yet made little progress amongst the assayers of Cornwall." The assays are conducted by the dry mode, and are somewhat analagous to the process of reduction on a large scale, and are nearly the same as those described by Price 60 years ago. The humid assay is the most exact, but it requires more skill and time. Ure gives several methods, (dry and wet.) The following, for a dry assay, is perhaps as good as any other: "A portion of the mixture, (prepared ore,) tried by the blow pipe, will show, by the garlic or sulphurous smell of its fumes, whether arsenic, sulphur, or both, be the mineralizers. In the latter case, which often occurs, 100 grains or 1000 grains of the ore are to be mixed with the one-half its weight of saw-dust, then imbued with oil, and heated moderately in a crucible, till all the arsenical fumes be dissipated. The residuum, being cooled and triturated, is to be exposed in a shallow earthen cup to a slow roasting heat till the sulphur and charcoal be burned away; what remains being ground and mixed with half its weight of calcined borax; one-twelfth its weight of lampblack, next, made into dough, with a few drops of oil, is to be pressed down into a crucible, which is to be covered with a luted lid, and to be subjected, in a powerful air furnace, first to a dull red heat, then to vivid ignition for 20 minutes. On cooling and breaking the crucible, a button of metallic copper will be obtained. Its color and malleability indicate pretty well the quality, as does its weight the relative value of the ore. It should be cupelled with lead, to ascertain if it contain silver or gold.

Ores of the oxide of copper are easily analyzed by solution in nitric acid, the addition of ammonia to separate the other metals, and precipitation by potash. The native carbonate is analyzed by calcining 100 grains, when the loss of weight will show the amount of water and carbonic acid; the latter may be found by expelling it from another 100 grains, by digestion in a grain weight of sulphuric acid. The copper is finally obtained in a metallic state, by plunging bars of zinc into the solution of the sulphate."

In the case of calcining, in heaps, in the open air—the ore is piled up in great masses in the form of truncated quadrangular pyramids, with the fuel at the bottom. In the centre of the mound is a wooden chimney, with charcoal at the bottom: an opening is left in the middle of each side, through which a current of air is admitted under the wooden chimney, by which means ignition is produced, when fire is thrown down the chimney, and combustion supported. This pyramid is covered with mortar, sods, &c., and hemispherical cavities are dug on the upper surface, for the purpose of receiving the sulphur which, during the roasting of ore, arrives liquified at the surface This method of roasting, which is practicable only when the ore is very sulphurous, is employed at Chessy, near Lyons, in France, and at Goslar, in the Lower Hartz. It lasts about six months, at the end of which period the sulphur ceases to escape. The mass is then left to cool, and when this is accomplished the ore is fit for smelting.

In the calcining operations, the volatile substances are mostly disengaged in a gaseous form; while the metals, owing to their strong affinity for oxygen, become oxidized. In melting, the earthy matters unite with these oxides, and form slags, which float on the surface of the molten metal.

Dr. Houghton, in his interesting report on the geology of Michigan, states that the carbonates of copper are the predominating ores of that metal found on the southern shores of Lake Superior. The ores from the same region lately exhibited to me by General Cunningham, U. S. agent of mineral lands, and other gentlemen, consisted of native copper, the red and black oxides, and the hydrosilicate, or *chrysocolla*, which, indeed, appears to be the predominating ore. According to Ure, it is green, or bluish green; specific gravity, 2.03 to 2.16; scratched by steel; is very brittle; affords water by heating, and blackens; is acted on by acids, and leaves a siliceous residuum. Solution becomes blue with ammonia. Its constituents are—silica, 26; oxide of copper, 50; water, 17; carbonic acid, 7. When pure, will yield 40 to 44 per cent. of metal. It is infusible at the blowpipe alone, but melts readily with borax. This description will apply to the specimens from Lake Superior; but I have never heard of its having been found elsewhere in sufficient quantities to be worked as an ore of copper. I see no reason, however, why it might not be easily reduced, using limestone or fluate of lime as a flux. The red oxides shown to me were not very rich, nor were they difficult of reduction. On the whole, I should think they might be worked profitably. The black oxides seemed to be much richer; but I could procure none for experiments.

The most certain indication of the progress of refining is found in a small sample of copper called a *montre*, which is taken from time to time from the bath on the heated end of a polished steel rod, plunged two or three inches deep into the bath, and then immersed in cold water. The montre is detached from the rod by a few smart blows of a hammer; and an opinion is formed of the degree of purity it has attained, from its density, colour, and polish. But these samples are never drawn till the drizzling of metallic globules ceases. The first montres withdrawn are thick; the surface is uniform, smooth, and of a red colour, resembling old copper coins; the interior is unequal, of a leaden colour, and dotted over with white and yellow specks. As the refining process advances, the small holes observed in the first montre gradually disappear, the outer surface becomes darker, the inner of a more uniform colour, and less and less marked with spots. Finally, when the metal has acquired the greatest degree of purity that this operation can give, the sample acquires a dark crimson appearance, tinged with a maroon shade; at the same time, the interior is of a deep and uniform red colour, unmixed with spots of any kind. In this state, which indicates the greatest possible degree of purity that copper can attain, it is flexible, and its fracture presents a close soft-grained appearance, of an obscure red colour.

The process must be stopped at the proper time, otherwise a small quantity of oxide of copper would be formed, rendering the metal hard, brittle, and incapable of perfect lamination.

Having ascertained by an examination of the last montre, that the copper has been sufficiently refined, the action of the bellows is stopped, the opening of the tuyere is closed, the tap is pierced, and the melted copper flows into the vessel prepared for its reception, from which the burning charcoal has been removed. When the receptacles are full, the tap-hole is reclosed with loam. When the surface of the metal which has been drawn off into the basins, is covered with a solid crust, it is slightly sprinkled with water, and when about two inches thick is raised up by means of hooks to let the drops drain off, and is then removed from the furnace. This operation is continued till all the copper has been taken off. These layers are called "*rosettes*." They are cooled suddenly by plunging them in water, when they assume a beautiful red colour, owing to the removal from their surface of a slight film of oxide of copper, by the volatilixation of the water. If the metal should be dipped into water before it has become completely solidified, dangerous explosions may happen; and if left too long cooling in the open air, it becomes peroxidized on the surface, and does not assume those beautiful colours which the protoxide imparts.

Each refining operation produces, in from 14 to 15 hours, when the hearth is new, and in from nine to ten hours when it is old, one and seven-tenths of a ton of fine copper, with the consumption of four-fifths of a ton of dry wood.

Care is taken, after each "drawing off," to close every fissure by which the exterior air might penetrate, in order to produce a gradual cooling of the furnace, and thus to avoid the cracking of the vitrified interior casing, which might be caused by a sudden refrigeration.

When it is deemed expedient to renew the hearth or the interior of the furnace, the old work is broken up, and the materials carefully washed, for the purpose of extracting the little metal which it always contains; small quantities of copper which have been sublimated, are also extracted from the flues and other parts of the furnace.

THE REFINING OF COPPER.

The process of refining copper is conducted in a reverbatory furnace, the hearth of which is composed of charbonnaile* or of quartz. The object of this operation is to evaporate all the volatile substance contained in the coarse copper, such as sulphur, arsenic, antimony, &c., and to oxidize and convert into scoriæ the fixed substances, such as iron, lead, &c., in such a manner as to allow the least possible waste of metalic copper. This operation is not very rigorous; but the small portion of gold or silver (which it sometimes contains) that will not oxidize is of no injury to the copper.

At the celebrated copper foundry of Seville, in Spain, where the very best quality of metal is produced, the following processes are observed:

Before heating the blast furnace, whether after its renewal or the reconstruction of its hearth, the whole interior is plastered with mortar of fire clay or pulverized brick, for the purpose of closing the cracks and preserving the vaults from the first impression of the fire

In arranging the charge, rows of fire brick are placed edgewise on the hearth of the furnace, and on the bricks are deposited successive layers of pigs of black or crude copper, crossing each other, with sufficient spaces between the pigs to permit the flame to circulate freely through and amongst the different layers, and to penetrate to the hearth, which it dries and heats sufficiently to produce a gradual fusion of the metal. Care is taken not to obstruct the vent near the flues, and not to place the pigs nearer than five or six inches to the walls of the furnace.

The pigs forming the first bed or layer are carefully placed, so that they may not, in falling, break upon the hearth.

The weight of the charge is proportioned to the capacity of the furnace, and so disposed that the level of the metalic vat, or bath, may be about one inch above the tuyere of the blast; for if it should be higher than this limit, the metal would attach itself to the tuyere, and thus obstruct the blast; and if it should be lower, the current of air would strike the surface of the molten metal imperfectly; in consequence of which, the refining process would be much retarded, by leaving the oxidation and volatilization of the foreign substances incomplete.

---

* This substance is composed of one-third refractory clay, one-third refractory sand, and one-third pulverized carbon, (in volume;) the whole moistened and thoroughly worked till it forms a consistent and homogeneous material, which vitrifies to a strong glassy mass by an intense heat.

The judicious management of the fire in a furnace which has been recently repaired consists in advancing the heat very slowly for the first three hours with very dry wood, to gradually deprive the hearth or charbonnaile of its moisture, and to soften the copper, so that it may fuse almost imperceptibly.

If, at the termination of these three hours of heating, it is noticed that the copper has become red, and that the bottom or hearth of the furnace has not yet lost its moisture, it will become proper to continue this slow combustion until every part of the furnace has acquired the same temperature with the copper. If, finally, this equilibrium of heat is not observed to take place, it will be necessary to suspend the fire for a short time, and to seal hermetically all the outlets or fissures of the furnace, in order to compel the copper to divide with it, its excess of temperature.

When the same degree of heat is thus produced in all parts of the furnace, the blast is forced; and at the end of seven or eight hours the metal begins to melt, and is soon after covered with a great quantity of scoriæ.

If it should be considered desirable to increase the intensity of the fire, it must be recollected that this object is not attained by overloading the grates with fuel; otherwise you obtain smoke only, which lowers rather than increases the temperature. The best guide to follow in this respect is the flame, which, when there is neither excess nor deficiency of combustible, fills the entire capacity of the furnace, and presents a beautifully intense red appearance.

When the bath has become fluid, and the scoriæ sufficiently liquid to give up the copper which they retain, they are removed with a long-handled rabble. If they appear refractory when subjected to a strong blast, they may be brought to a suitable vitrification by mixing limestone or calcareous clay with them for a flux. If, on the contrary, they become too fluid to be raked off, which is often the case, they may be thickened by throwing in refractory materials similar to those of which the hearth is composed. After having freed the bath of all impurities, the bellows are put in action.

Soon after the metal is melted, charcoal is ignited in three large iron vessels, lined with loam, to prepare them for the reception of the copper, which is to be converted into rosettes.

Ordinarily, soon after the bellows are put in motion, the evaporation of mineral substances is so abundant as to produce ebullition in the bath; some of the drops rise even to the vault of the furnace, while others escape from the doors, and fall condensed in a drizzling rain of small spherical globules. When this phenomenon appears, the refining goes on well; when it disappears, the operation is drawing to an end.